D1499336

NETWORK MODELING AND SIMULATION

NETWORK MODELING AND SIMULATION
A PRACTICAL PERSPECTIVE

Mohsen Guizani
Kuwait University, Kuwait

Ammar Rayes
Cisco Systems, USA

Bilal Khan
City University of New York, USA

Ala Al-Fuqaha
Western Michigan University, USA

A John Wiley and Sons, Ltd, Publication

Registered office
John Wiley & Sons Ltd, The Atrium, Southern Gate, Chichester, West Sussex, PO19 8SQ, United Kingdom

For details of our global editorial offices, for customer services and for information about how to apply for permission to reuse the copyright material in this book please see our website at www.wiley.com.

Library of Congress Cataloging-in-Publication Data

Network modeling and simulation : a practical perspective / M. Guizani ... [et al.].
    p. cm.
  Includes bibliographical references and index.
  ISBN 978-0-470-03587-0 (cloth)
  1. Simulation methods. 2. Mathematical models. 3. Network analysis (Planning)–Mathematics. I. Guizani, Mohsen.
  T57.62.N48 2010
  003'.3–dc22

                             2009038749

A catalogue record for this book is available from the British Library.

ISBN 978-0-470-03587-0 (H/B)

Set in 11/13 Times Roman by Thomson Digital, Noida, India
Printed and bound in Great Britain by CPI Antony Rowe, Chippenham, Wiltshire

Contents

Preface

Networking technologies are growing more complex by the day. So, one of the most important requirements for assuring the correct operation and rendering of the promised service to demanding customers is to make sure that the network is robust. To assure that the network is designed properly to support all these demands before being operational, one should use the correct means to model and simulate the design and carry out enough experimentation. So, the process of building good simulation models is extremely important in such environments, which led to the idea of writing this book.

In this book, we chose to introduce generic simulation concepts and frameworks in the earlier chapters and avoid creating examples that tie the concepts to a specific industry or a certain tool. In later chapters, we provide examples that tie the simulation concepts and frameworks presented in the earlier chapters to computer and telecommunications networks. We believe that this will help illustrate the process of mapping the generic simulation concepts to a specific industry.

Therefore, we have concentrated on the core concepts of systems simulation and modeling. We also focused on equipping the reader with the tools and strategies needed to build simulation models and solutions from the ground up rather than provide solutions to specific problems. In addition, we presented code examples to illustrate the implementation process of commonly encountered simulation tasks.

The following provides a chapter-by-chapter breakdown of this book's material.

Chapter 1 introduces the foundations of modeling and simulation, and emphasizes their importance. The chapter surveys the different approaches to modeling that are used in practice and discusses at a high level the methodology that should be followed when executing a modeling project.

In Chapter 2, we assemble a basic discrete event simulator in Java. The framework is not very large (less than 250 lines of code, across three classes) and yet it is extremely powerful. We deduce the design of the simulation framework (together with its code). Then, we discuss a few "toy examples" as a tutorial on how to write applications over the framework.

In Chapter 3, we turn to a case study that illustrates how to conduct large discrete event simulations using the framework designed in Chapter 2. We then design and develop a simulation of a system that will generate malware antivirus signatures using an untrusted multi-domain community of honeypots (as a practical example encountered usually in today's networks).

Chapter 4 introduces the well-known Monte Carlo simulation technique. The technique applies to both deterministic and probabilistic models to study properties of stable systems that are in equilibrium. A random number generator is used by Monte Carlo to simulate a performance measure drawn from a population with appropriate statistical properties. The Monte Carlo algorithm is based on the law of large numbers with the promise that the mean value of a large number of samples from a given space will approximate the actual mean value of such a space.

Chapter 5 expands upon the concepts introduced in earlier chapters and applies them to the area of network modeling and simulation. Different applications of modeling and simulation in the design and optimization of networked environments are discussed. We introduce the network modeling project life cycle and expose the reader to some of the particular considerations when modeling network infrastructures. Finally, the chapter attempts to describe applications of network modeling within the linkage between network modeling and business requirements.

In Chapter 6, we define a framework that will allow for modular specification and assembly of dataflow processing modules within a single device. We call this framework the Component Architecture for Simulating Network Objects (CASiNO). A discussion on how to use CASiNO and code its components is presented in some detail.

Then, in Chapter 7, we study a set of statistical distributions that could be used in simulation as well as a set of random number generation techniques.

In Chapter 8, we create some useful network simulation elements that will serve as building blocks in the network structures that we consider in the context of queuing theory in Chapter 9.

Chapter 9 presents a brief discussion on several topics in queuing theory. In the first part, we cover the basic concepts and results, whereas in the second part we discuss specific cases that arise frequently in practice. Whenever possible, there are code samples implemented using the CASiNO framework (developed in Chapter 6), the SimJava Package, and the MATLAB package.

Chapter 10 elaborates on the importance of data collection as a phase within the network modeling project life cycle. It lists the different data types that need to be collected to support network modeling projects, and how to collect the data, choose the right distribution, and validate the correctness of one's choice.

Chapter 11 presents traffic models used to simulate network traffic loads. The models are divided into two main categories: models which exhibit long-range dependencies or self-similarities and Markovian models that exhibit only short-range dependence. Then, an overview of some of the commonly used global optimization techniques to solve constrained and unconstrained optimization problems are presented. These techniques are inspired by the social behaviors of birds, natural selection and survival of the fittest, and the metal annealing process as well as the fact of trying to simulate such behaviors.

Finally, we hope that this book will help the reader to understand the code implementation of a simulation system from the ground up. To that end, we have built a new simulation tool from scratch called "CASiNO." We have also treated all the examples in a step-by-step fashion to keep the user aware of what is happening and how to model a system correctly. So, we hope that this book will give a different flavor to modeling and simulation in general and to that of network modeling and simulation in particular.

Mohsen Guizani
Ammar Rayes
Bilal Khan
Ala Al-Fuqaha

Acknowledgments

We realize that this work would not have been a reality without the support of so many people around us. First, we would like to express our gratitude to Cisco Systems for partly supporting the research work that has contributed to this book. In particular, thanks to Mala Anand, VP of Cisco's Smart Services, and Jim McDonnell, Senior Director of Cisco's Smart Services. The Cisco research Grant to Western Michigan University was the nucleus of this project. Therefore, we are grateful for Cisco's research support for the last three years. Also, special thanks to the students of the Computer Science Department at Western Michigan University who worked on Cisco's research project and eventually played an important role in making this project a success. Mohsen Guizani is grateful to Kuwait University for its support in this project. Ammar Rayes is thankful to his department at Cisco Systems for encouragement and support. Ala Al-Fuqaha appreciates the support of Western Michigan University. Bilal Khan acknowledges collaboration with Hank Dardy and the Center for Computational Science at the US Naval Research Laboratory. He also thanks The John Jay College at CUNY for its support, and the Secure Wireless Ad-hoc Network (SWAN) Lab and Social Network Research Group (SNRG) for providing the context for ongoing inquiry into network systems through simulation.

The authors would like also to thank the John Wiley & Sons, Ltd team of Mark Hammond, Sarah Tilley, and Sophia Travis for their patience and understanding throughout this project.

Last but not least, the authors are grateful to their families. Mohsen Guizani is indebted to all members of his family, his brothers and sisters, his wife Saida, and his children: Nadra, Fatma, Maher, Zainab, Sara, and Safa. Ammar Rayes would like to express thanks to his wife Rana, and his kids: Raneem, Merna, Tina, and Sami. Ala Al-Fuqaha is indebted to his parents as well his wife Diana and his kids Jana and Issam.

1

Basic Concepts and Techniques

This chapter introduces the foundations of modeling and simulation, and elucidates their importance. We survey the different approaches to modeling that are used in practice and discuss at a high level the methodology that should be followed when executing a modeling project.

1.1 Why is Simulation Important?

Simulation is the imitation of a real-world system through a computational re-enactment of its behavior according to the rules described in a mathematical model.[1] Simulation serves to imitate a real system or process. The act of simulating a system generally entails considering a limited number of key characteristics and behaviors within the physical or abstract system of interest, which is otherwise infinitely complex and detailed. A simulation allows us to examine the system's behavior under different scenarios, which are assessed by re-enactment within a virtual computational world. Simulation can be used, among other things, to identify bottlenecks in a process, provide a safe and relatively cheaper (in term of both cost and time) test bed to evaluate the side effects, and optimize the performance of the system—all before realizing these systems in the physical world.

Early in the twentieth century, modeling and simulation played only a minor role in the system design process. Having few alternatives, engineers moved straight from paper designs to production so they could test their designs. For example, when Howard Hughes wanted to build a new aircraft, he never knew if it would fly until it was

---

[1] Alan Turing used the term *simulation* to refer to what happens when a digital computer runs a finite state machine (FSM) that describes the state transitions, inputs, and outputs of a subject discrete-state system. In this case, the computer simulates the subject system by executing a corresponding FSM.

actually built. Redesigns meant starting from scratch and were very costly. After World War II, as design and production processes became much more expensive, industrial companies like Hughes Aircraft collapsed due to massive losses. Clearly, new alternatives were needed. Two events initiated the trend toward modeling and simulation: the advent of computers; and NASA's space program. Computing machinery enabled large-scale calculations to be performed quickly. This was essential for the space program, where projections of launch and re-entry were critical. The simulations performed by NASA saved lives and millions of dollars that would have been lost through conventional rocket testing. Today, modeling and simulation are used extensively. They are used not just to find if a given system design works, but to discover a system design that works best. More importantly, modeling and simulation are often used as an inexpensive technique to perform exception and "what-if" analyses, especially when the cost would be prohibitive when using the actual system. They are also used as a reasonable means to carry out stress testing under exceedingly elevated volumes of input data.

Japanese companies use modeling and simulation to improve quality, and often spend more than 50% of their design time in this phase. The rest of the world is only now beginning to emulate this procedure. Many American companies now participate in rapid prototyping, where computer models and simulations are used to quickly design and test product concept ideas before committing resources to real-world in-depth designs.

Today, simulation is used in many contexts, including the modeling of natural systems in order to gain insight into their functioning. Key issues in simulation include acquisition of valid source information about the referent, selection of key characteristics and behaviors, the use of simplifying approximations and assumptions within the simulation, and fidelity and validity of the simulation outcomes. Simulation enables goal-directed experimentation with dynamical systems, i.e., systems with time-dependent behavior. It has become an important technology, and is widely used in many scientific research areas.

In addition, modeling and simulation are essential stages in the engineering design and problem-solving process and are undertaken before a physical prototype is built. Engineers use computers to draw physical structures and to make mathematical models that simulate the operation of a device or technique. The modeling and simulation phases are often the longest part of the engineering design process. When starting this phase, engineers keep several goals in mind:

- Does the product/problem meet its specifications?
- What are the limits of the product/problem?
- Do alternative designs/solutions work better?

The modeling and simulation phases usually go through several iterations as engineers test various designs to create the best product or the best solution to a

Table 1.1 Examples of simulation usage in training and education

Usage of simulation	Examples
Training to enhance motor and operational skills (and associated decision-making skills)	• Virtual simulation which uses virtual equipment and real people (human-in-the-loop) in a simulation study • Aircraft simulator for pilot training • Augmented reality simulation (such as in-flight pilot training with additional artificial intelligence aircraft) • Virtual body for medicine • Nuclear reactor simulator • Power plant simulator • Simulators for the selection of operators (such as pilots) • Live simulation (use of simulated weapons along with real equipment and people)
Training to enhance decision-making skills	• Constructive simulation (war game simulation) • Simulation for operations other than war (non-article 5 operations, in NATO terminology): peace support operations; conflict management (between individuals, groups, nations) • Business game simulations • Agent-based simulations • Holonic agent simulations that aim to explore benefits of cooperation between individuals, companies (mergers), and nations
Education	• Simulation for the teaching/learning of dynamic systems (which may have trajectory and/or structural behavior): simulation of adaptive systems, time-varying systems, evolutionary systems, etc.

problem. Table 1.1 provides a summary of some typical uses of simulation modeling in academia and industry to provide students and professionals with the right skill set.

Simulation provides a method for checking one's understanding of the world and helps in producing better results faster. A simulation environment like MATLAB is an important tool that one can use to:

• Predict the course and results of certain actions.
• Understand why observed events occur.
• Identify problem areas before implementation.
• Explore the effects of modifications.

- Confirm that all variables are known.
- Evaluate ideas and identify inefficiencies.
- Gain insight and stimulate creative thinking.
- Communicate the integrity and feasibility of one's plans.

One can use simulation when the analytical solution does not exist, is too complicated, or requires more computational time than the simulation. Simulation should not be used in the following cases:

- The simulation requires months or years of CPU time. In this scenario, it is probably not feasible to run simulations.
- The analytical solution exists and is simple. In this scenario, it is easier to use the analytical solution to solve the problem rather than use simulation (unless one wants to relax some assumptions or compare the analytical solution to the simulation).

1.2 What is a Model?

A computer model is a computer program that attempts to simulate an abstract model of a particular system. Computer models can be classified according to several orthogonal binary criteria including:

- **Continuous state or discrete state:** If the state variables of the system can assume any value, then the system is modeled using a *continuous-state model*. On the other hand, a model in which the state variables can assume only discrete values is called a *discrete-state model*. Discrete-state models can be continuous- or discrete-time models.
- **Continuous-time or discrete-time models:** If the state variables of the system can change their values at any instant in time, then the system can be modeled by a *continuous-time model*. On the other hand, a model in which the state variables can change their values only at discrete instants of time is called a *discrete-time model*.

A continuous simulation uses differential equations (either partial or ordinary), implemented numerically. Periodically, the simulation program solves all the equations, and uses the numbers to change the state and output of the simulation. Most flight and racing-car simulations are of this type. It may also be used to simulate electric circuits. Originally, these kinds of simulations were actually implemented on analog computers, where the differential equations could be represented directly by various electrical components such as op-amps. By the late 1980s, however, most "analog" simulations were run on conventional digital computers that emulated the behavior of an analog computer.

A discrete-time event-driven simulation (DES) manages events in time [3]. In this type of simulation, the simulator maintains a queue of events sorted by the simulated time they should occur. The simulator reads the queue and triggers new events as each event is processed. It is not important to execute the simulation in real time. It is often more important to be able to access the data produced by the simulation, to discover logic defects in the design or the sequence of events. Most computer, logic test and fault-tree simulations are of this type.

A special type of discrete simulation that does not rely on a model with an underlying equation, but can nonetheless be represented formally, is *agent-based simulation*. In agent-based simulation, the individual entities (such as molecules, cells, trees, or consumers) in the model are represented directly (rather than by their density or concentration) and possess an internal state and set of behaviors or rules which determine how the agent's state is updated from one time step to the next.

- **Deterministic or probabilistic:** In a *deterministic model*, repetition of the same input will always yield the same output. On the other hand, in a *probabilistic model*, repetition of the same input may lead to different outputs. Probabilistic models use random number generators to model the chance or random events; they are also called *Monte Carlo (MC) simulations* (this is discussed in detail in Chapter 4). *Chaotic models* constitute a special case of deterministic continuous-state models, in which the output is extremely sensitive to input values.
- **Linear or nonlinear models:** If the outputs of the model are linearly related to the inputs, then the model is called a *linear model*. However, if the outputs are not linear functions of the inputs, then the model is called a *nonlinear model*.
- **Open or closed models:** If the model has one or more external inputs, then the model is called an *open model*. On the other hand, the model is called a *closed model* if it has no external inputs at all.
- **Stable or unstable models:** If the dynamic behavior of the model comes to a steady state with time, then the model is called *stable*. Models that do not come to a steady state with time are called *unstable models*. Note that stability refers to time, while the notion of chaos is related to behavioral sensitivities to the input parameters.
- **Local or distributed:** If the model executes only on a single computer then it is called a *local model*. *Distributed models*, on the other hand, run on a network of interconnected computers, possibly in a wide area, over the Internet. Simulations dispersed across multiple host computers are often referred to as *distributed simulations*. There are several military standards for distributed simulation, including Aggregate Level Simulation Protocol (ALSP), Distributed Interactive Simulation (DIS), and the High Level Architecture (HLA).

1.2.1 Modeling and System Terminology

The following are some of the important terms that are used in simulation modeling:

- **State variables:** The variables whose values describe the state of the system. They characterize an attribute in the system such as level of stock in an inventory or number of jobs waiting for processing. In the case where the simulation is interrupted, it can be completed by assigning to the state variables the values they held before interruption of the simulation.
- **Event:** An occurrence at a point in time, which may change the state of the system, e.g., the arrival of a customer or start of work on a job.
- **Entity:** An object that passes through the system, such as cars at an intersection or orders in a factory. Often events (e.g., arrival) are associated with interactions between one or more entities (e.g., customer and store), with each entity having its own state variables (e.g., customer's cash and store's inventory).
- **Queue:** A queue is a linearly ordered list, e.g., a physical queue of people, a task list, a buffer of finished goods waiting for transportation. In short, a place where entities are waiting for something to happen for some coherent reason.
- **Creating:** Causing the arrival of a new entity to the system at some point in time. Its dual process will be referred to as *killing* (the departure of a state).
- **Scheduling:** The act of assigning a new future event to an existing entity.
- **Random variable:** A random variable is a quantity that is uncertain, such as the arrival time between two incoming flights or the number of defective parts in a shipment.
- **Random variate:** A random variate is an artificially generated random variable.
- **Distribution:** A distribution is the mathematical law that governs the probabilistic features of a random variable. Examples of frequently used distributions are discussed in Chapter 7.

1.2.2 Example of a Model: Electric Car Battery Charging Station

Let us discuss the process of building a simulation of a charging station for electric cars at a single station served by a single operative. We assume that the arrival of electric cars as well as their service times are random. First, we have to identify:

- **Entities:** The entities are of three types: cars (multiple instances), the charging station server (single instance), and the traffic generator (single instance) which creates cars and sends them to the battery charging station.
- **States:** The charging station has a server that is in either idle or free mode. Each car C_i has a battery of some capacity K_i and some amount of power P_i present within it. The traffic generator has parameters that specify the distribution of random inter-creation times of cars.

- **Events:** One kind of event is "service car," which is an instruction to the server to charge the next car's battery. Another kind of event is the arrival event, which is sent by the traffic generator to the charging station to inform it that a car has arrived. Finally, there is the "create car event," which is sent by the traffic generator to itself to trigger the creation of a car.
- **Queue:** The queue of cars Q in front of the charging station.
- **Random realizations:** Car inter-creation times, car battery capacities, battery charge levels (and, consequently, car service times), etc.
- **Distributions:** Assume uniform distributions for the car inter-creation times, battery capacities, and charge levels.

Next, we specify what to do at each event.

The traffic generator sends itself a car creation event. Upon receipt of this event, it creates a car with a battery capacity that is randomly chosen uniformly in the interval $[0...F]$, and the battery is set to be L% full, where L is randomly chosen in the interval $[0...100]$. The traffic generator encapsulates the car in a car arrival event and sends the event to the charging station. It also sends itself another car creation event for some time (chosen uniformly from an interval $[0...D]$ minutes) in the future.

When a car arrival event is received at the charging station, the car is added to the end of the queue. If (at the moment just prior to this occurrence) the server was in free mode, and the queue transitioned from empty to non-empty, then the server is sent a service car event.

The server, upon receiving a service car event, checks if Q is empty. If it is, the server goes into idle mode. If Q is not empty, the server puts itself in a busy mode, removes one car C_i from the head of the queue Q, and charges its battery. In addition, the server calculates how long it will take to charge the car's battery as $(K_i - P_i)/M$, where M is the maximum rate at which power can be charged. The server then schedules another service car event to be sent to itself after this amount of time has elapsed.

As this simulation process runs, the program records interesting facts about the system, such as the number of cars in Q as time progresses and the fraction of time that the server is in idle versus busy modes. The maximum size of Q over the simulation's duration, and the average percentage of time the server is busy, are two examples of performance metrics that we might be interested in.

Some initiation is required for the simulation; specifically we need to know the values of D (which governs car inter-creation intervals), F (which governs battery size), and M (which is the maximum charging rate (power flow rate)). Thus D, F, and M are system parameters of the simulation. The values of these parameters must be set to reasonable values for the simulation to produce interpretable results. The choice of values requires detailed domain expertise—in this case, perhaps through consultation with experts, or by statistical analysis of real trace data from a real-world charging station.

Once these event-driven behaviors are defined, they can be translated into code. This is easy with an appropriate library that has subroutines for the creation, scheduling, and proper timing of events, queue manipulations, random variable generation, and collecting statistics. This charging station system will be implemented as a case study in Chapter 4, using a discrete-event simulation framework that will be designed and implemented in Chapter 2.

The performance measures (maximum size of Q over the simulation's duration, and the average percentage of time the server is busy) depend heavily on random choices made in the course of the simulation. Thus, the system is non-deterministic and the performance measures are themselves random variables. The distributions of these random variables using techniques such as Monte Carlo simulation will be discussed in Chapter 4.

1.3 Performance Evaluation Techniques

Modeling deals with the representation of interconnected subsystems and subprocesses with the eventual objective of obtaining an estimate of an aggregate systemic property, or *performance measure.* The mechanism by which one can obtain this estimate (1) efficiently and (2) with confidence is the subject of *performance evaluation.* In this section, we will lay out the methodological issues that lie therein.

A *performance evaluation technique* (PET) is a set of assumptions and analytical processes (applied in the context of a simulation), whose purpose is the efficient estimation of some performance measure. PETs fall into two broad classes: (1) direct measurement and (2) modeling:

1. **Direct measurement:** The most obvious technique to evaluate the performance of a system. However, limitations exist in the available traffic measurements because of the following reasons:
   (a) The direct measurement of a system is only available with operational systems. Direct measurement is not feasible with systems under design and development.
   (b) Direct measurement of the system may affect the measured system while obtaining the required data. This may lead to false measurements of the measured system.
   (c) It may not be practical to measure directly the level of end-to-end performance sustained on all paths of the system. Hence, to obtain the performance objectives, analytical models have to be set up to convert the raw measurements into meaningful performance measures.
2. **Modeling:** During the design and development phases of a system, modeling can be used to estimate the performance measures to be obtained when the system is

implemented. Modeling can be used to evaluate the performance of a working system, especially after the system undergoes some modifications. Modeling does not affect the measured system as it does in the direct measurement technique and it can be used during the design phases. However, modeling suffers from the following problems:

(a) The problem of *system abstraction*. This may lead to analyzing a model that does not represent the real system under evaluation.
(b) Problems in representing the workload of the system.
(c) Problems in obtaining performance measures for the model and mapping the results back to the real system.

Needless to say, the direct measurement of physical systems is *not* the subject of this book. Even within the modeling, only a sub-branch is considered here. To see where this sub-branch lies, it should be noted that at the heart of every model is its formal description. This can almost always be given in terms of a collection of interconnected queues. Once a system has been modeled as a collection of interconnected queues, there are two broadly defined approaches for determining actual performance measures: (1) analytical modeling and (2) simulation modeling. Analytical modeling of such a system seeks to deduce the parameters through mathematical derivation, and so leads into the domain of *queuing theory*—a fascinating mathematical subject for which many texts are already available. In contrast, simulation modeling (which is the subject of this book) determines performance measures by making *direct measurements of a simplified virtual system that has been created through computational means.*

In simulation modeling, there are few PETs that can be applied generally, though many PETs are based on modifications of the so-called Monte Carlo method that will be discussed in depth in Chapter 4. More frequently, each PET typically has to be designed on a case-by-case basis given the system at hand. The process of defining a PET involves:

1. Assumptions or simplifications of the properties of the atomic elements within a system and the logic underlying their evolution through interactions.[2]
2. Statistical assumptions on the properties of non-deterministic aspects of the system, e.g., its constituent waveforms.
3. Statistical techniques for aggregating performance measure estimates obtained from disparate simulations of different system configurations.[3]

---

[2] Note that this implies that a PET may be intertwined with the process of model construction, and because of this, certain PETs imply or require departure from an established model.

[3] At the simplest level, we can consider Monte Carlo simulation to be a PET. It satisfies our efficiency objective to the extent that the computations associated with missing or simplified operations reduce the overall computational burden.

To illustrate, let us consider an example of a PET methodology that might be applied in the study of a high data rate system over slowly fading channels. Without going into too many details, such a system's performance depends upon two random processes with widely different rates of change in time. One of the processes is "fast" in terms of its dynamism while the other is "slow." For example, the fast process in this case might be thermal noise, while the slow process might be signal fading. If the latter process is sufficiently slow relative to the first, then it can be approximated as being considered fixed with respect to the fast process. In such a system, the evolution of a received waveform could be simulated, by considering a sequence of time segments over each of which the state of the slow process is different, but fixed. We might simulate such a sequence of segments, and view each segment as being an experiment (conditioned on the state of the slow process) and obtain conditional performance metrics. The final performance measure estimate might then be taken as the average over these conditional estimates. This narrative illustrates the process of selecting and specifying a PET.

1.3.1 Example of Electric Car Battery Charging Station

In Section 1.2.1, a model of a charging station was described. What might the associated performance measures be when we simulate the model? There are many possible interesting aggregate performance measures we could consider, such as the number of cars in Q as time progresses and the fraction of time when the server is in idle versus busy modes.

1.3.1.1 Practical Considerations

Simulation is, in some sense, an act of pretending that one is dealing with a real object when actually one is working with an imitation of reality. It may be worthwhile to consider when such pretense is warranted, and the benefits/dangers of engaging in it.

1.3.1.2 Why Use Models?

Models reduce cost using different simplifying assumptions. A spacecraft simulator on a certain computer (or simulator system) is also a computer model of some aspects of the spacecraft. It shows on the computer screen the controls and functions that the spacecraft user is supposed to see when using the spacecraft. To train individuals using a simulator is more convenient, safer, and cheaper than a real spacecraft that usually cost billions of dollars. It is prohibitive (and impossible) to use a real spacecraft as a training system. Therefore, industry, commerce, and the military often use models rather than real experiments. Real experiments can be as costly and dangerous as real systems so, provided that models are adequate descriptions of reality, they can give results that are closer to those achieved by a real system analysis.

1.3.1.3 When to Use Simulations?

Simulation is used in many situations, such as:

1. When the analytical model/solution is not possible or feasible. In such cases, experts resort to simulations.
2. Many times, simulation results are used to verify analytical solutions in order to make sure that the system is modeled correctly using analytical approaches.
3. *Dynamic systems*, which involve randomness and change of state with time. An example is our electric car charging station where cars come and go unpredictably to charge their batteries. In such systems, it is difficult (sometimes impossible) to predict exactly what time the next car should arrive at the station.
4. *Complex dynamic systems*, which are so complex that when analyzed theoretically will require too many simplifications. In such cases, it is not possible to study the system and analyze it analytically. Therefore, simulation is the best approach to study the behavior of such a complex system.

1.3.1.4 How to Simulate?

Suppose one is interested in studying the performance of an electric car charging station (the example treated above). The behavior of this system may be described graphically by plotting the number of cars in the charging station and the state of the system. Every time a car arrives, the graph increases by one unit, while a departing car causes the graph to drop one unit. This graph, also called a *sample path*, could be obtained from observation of a real electric car charging station, but could also be constructed artificially. Such artificial construction and the analysis of the resulting sample path (or more sample paths in more complex cases) constitute the simulation process.

1.3.1.5 How Is Simulation Performed?

Simulations may be performed manually (on paper for instance). More frequently, however, the system model is written either as a computer program or as some kind of input to simulation software, as shown in Figure 1.1.

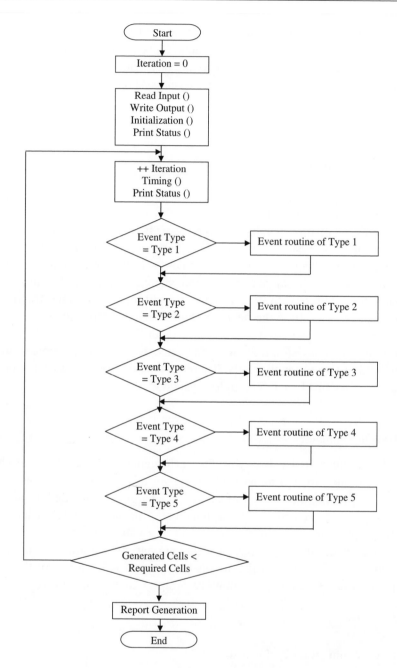

Figure 1.1   Overall structure of a simulation program

1.3.2 Common Pitfalls

In the following, we discuss the most common modeling and programming mistakes that lead to inaccurate or misleading simulation results:

1. **Inappropriate level of detail:** Normally, analytical modeling is carried out after some simplifications are adopted, since, without such simplifications, analysis tends to be intractable. In contrast, in simulation modeling, it is often tempting to include a very high level of simulation detail, since the approach is computational and not analytical. Such a decision is not always recommended, however, since much more time is needed to develop and run the computations. Even more importantly, having a model that is too detailed introduces a large number of interdependent parameters, whose influence on system performance becomes difficult to determine and isolate. For example, suppose we made our electric car charging station model include whether or not the server had a conversation with a passenger in the car, changed one of the car tires, and/or washed the car. In addition, each car had state variables concerning its battery charging level, passenger's gender, and battery type/size. If this model is used, it should contain details of this dynamic system. In this case, the simulation will get more complicated since it should contain details of the model that will make it so precise. But is this the best way to deal with the system? Is it worth including these details in the simulation model? Will the simulation results be better keeping the modeling details or it is better to leave them out since they will not add any necessary improvements to the results? Even if they do, will that be worth the added complexity to the system?

2. **Improper selection of the programming language:** The choice of programming language to be used in implementing the simulation model greatly affects the development time of the model. Special-purpose simulation languages require less time for development, while general-purpose languages are more efficient during the run time of the simulation program.

3. **Unverified models:** Simulation models are generally very large computer programs. These programs can contain several programming (or, even worse, logical) errors. If these programs are not verified to be correct, they will lead to misleading results and conclusions and invalid performance evaluation of the system.

4. **Invalid models:** Simulation programs may have no errors, but they may not represent the behavior of the real system that we want to evaluate. Hence, domain experts must verify the simulation assumptions codified by the mathematical model.

5. **Improperly selected initial conditions:** The initial conditions and parameter values used by a simulation program normally do not reflect the system's behavior in the steady state. Poorly selected initial conditions can lead to convergence to

atypical steady states, suggesting misleading conclusions about the system. Thus, domain experts must verify the initial conditions and parameters before they are used within the simulation program.

6. **Short run times:** System analysts may try to save time by running simulation programs for short periods. Short runs lead to false results that are strongly dependent on the initial conditions and thus do not reflect the true performance of the system in typical steady-state conditions.

7. **Poor random number generators:** Random number generators employed in simulation programs can greatly affect simulation results. It is important to use a random number generator that has been extensively tested and analyzed. The seeds that are supplied to the random number generator should also be selected to ensure that there is no correlation between the different processes in the system.

8. **Inadequate time estimate:** Most simulation models fail to give an adequate estimate of time needed for the development and implementation of the simulation model. The development, implementation, and testing of a simulation model require a great deal of time and effort.

9. **No achievable goals:** Simulation models may fail because no specific, achievable goals are set before beginning the process of developing the simulation model.

10. **Incomplete mix of essential skills:** For a simulation project to be successful, it should employ personnel who have different types of skills, such as project leaders and programmers, people who have skills in statistics and modeling, and people who have a good domain knowledge and experience with the actual system being modeled.

11. **Inadequate level of user participation:** Simulation models that are developed without end user's participation are usually not successful. Regular meetings among end users, system developers, and system implementers are important for a successful simulation program.

12. **Inability to manage the simulation project:** Most simulation projects are extremely complex, so it is important to employ software engineering tools to keep track of the progress and functionality of a project.

1.3.3 Types of Simulation Techniques

The most important types of simulations described in the literature that are of special importance to engineers are:

1. **Emulation:** The process of designing and building hardware or firmware (i.e., prototype) that imitates the functionality of the real system.

2. **Monte Carlo simulation:** Any simulation that has no time axis. Monte Carlo simulation is used to model probabilistic phenomena that do not change with time,

or to evaluate non-probabilistic expressions using probabilistic techniques. This kind of simulation will be discussed in greater detail in Chapter 4.

3. **Trace-driven simulation:** Any simulation that uses an ordered list of real-world events as input.

4. **Continuous-event simulation:** In some systems, the state changes occur all the time, not merely at discrete times. For example, the water level in a reservoir with given in- and outflows may change all the time. In such cases "continuous simulation" is more appropriate, although discrete-event simulation can serve as an approximation.

5. **Discrete-event simulation:** A discrete-event simulation is characterized by two features: (1) within any interval of time, one can find a subinterval in which no event occurs and no state variables change; (2) the number of events is finite. All discrete-event simulations have the following components:

   (a) *Event queue*: A list that contains all the events waiting to happen (in the future). The implementation of the event list and the functions to be performed on it can significantly affect the efficiency of the simulation program.

   (b) *Simulation clock*: A global variable that represents the simulated time. Simulation time can be advanced by *time-driven* or *event-driven* methods. In the time-driven approach, time is divided into constant, small increments, and then events occurring within each increment are checked. In the event-driven approach, on the other hand, time is incremented to the time of the next imminent event. This event is processed and then the simulation clock is incremented again to the time of the next imminent event, and so on. This latter approach is the one that is generally used in computer simulations.

   (c) *State variables*: Variables that together completely describe the state of the system.

   (d) *Event routines*: Routines that handle the occurrence of events. If an event occurs, its corresponding event routine is executed to update the state variables and the event queue appropriately.

   (e) *Input routine*: The routine that gets the input parameters from the user and supplies them to the model.

   (f) *Report generation routine*: The routine responsible for calculating results and printing them out to the end user.

   (g) *Initialization routine*: The routine responsible for initializing the values of the various state variables, global variables, and statistical variables at the beginning of the simulation program.

   (h) *Main program*: The program where the other routines are called. The main program calls the initialization routine; the input routine executes various iterations, finally calls the report generation routine, and terminates the simulation. Figure 1.1 shows the overall structure that should be followed to implement a simulation program.

1.4 Development of Systems Simulation

Discrete-event systems are dynamic systems that evolve in time by the occurrence of events at possibly irregular time intervals. Examples include traffic systems, flexible manufacturing systems, computer communications systems, production lines, coherent lifetime systems, and flow networks. Most of these systems can be modeled in terms of discrete events whose occurrence causes the system to change state. In designing, analyzing, and simulating such complex systems, one is interested not only in performance evaluation, but also in analysis of the sensitivity of the system to design parameters and optimal selection of parameter values.

A typical stochastic system has a large number of control parameters, each of which can have a significant impact on the performance of the system. An overarching objective of simulation is to determine the relationship between system behavior and input parameter values, and to estimate the relative importance of these parameters and their relationships to one another as mediated by the system itself. The technique by which this information is deduced is termed *sensitivity analysis*. The methodology is to apply small perturbations to the nominal values of input parameters and observe the effects on system performance measures. For systems simulation, variations of the input parameter values cannot be made infinitely small (since this would then require an infinite number of simulations to be conducted). The *sensitivity* (of the performance measure with respect to an input parameter) is taken to be an approximation of the partial derivative of the performance measure with respect to the parameter value.

The development process of system simulation involves some or all of the following stages:

- **Problem formulation:** This involves identifying the controllable and uncontrollable inputs (see Figure 1.2), identifying constraints on the decision variables, defining a measure of performance (i.e., an objective function), and developing a preliminary model structure to interrelate the system inputs to the values of the performance measure.

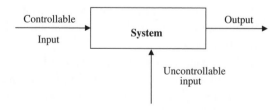

Figure 1.2   System block diagram

- **Data collection and analysis:** Decide *what* data to collect about the real system, and *how much* to collect. This decision is a tradeoff between cost and accuracy.
- **Simulation model development:** Acquire sufficient understanding of the system to develop an appropriate conceptual, logical model of the entities and their states, as well as the events codifying the interactions between entities (and time). This is the heart of simulation design.
- **Model validation, verification, and calibration:** In general, verification focuses on the internal consistency of a model, while validation is concerned with the correspondence between the model and the reality. The term *validation* is applied to those processes that seek to determine whether or not a simulation is correct with respect to the "real" system. More precisely, validation is concerned with the question "are we building the right system?" Verification, on the other hand, seeks to answer the question "are we building the system right?" Verification checks that the implementation of the simulation model (program) corresponds to the model. Validation checks that the model corresponds to reality. Finally, calibration checks that the data generated by the simulation matches real (observed) data.
- **Validation:** The process of comparing the model's output to the behavior of the phenomenon. In other words, comparing model execution to reality (physical or otherwise). This process and its role in simulation are described in Figure 1.3.
- **Verification:** The process of comparing the computer code to the model to ensure that the code is a correct implementation of the model.
- **Calibration:** The process of parameter estimation for a model. Calibration is a tweaking/tuning of existing parameters which usually does not involve the introduction of new ones, changing the model structure. In the context of optimization,

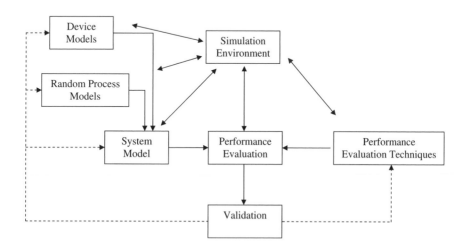

Figure 1.3 Validation process

calibration is an optimization procedure involved in system identification or during the experimental design.

- **Input and output analysis:** Discrete-event simulation models typically have stochastic components that mimic the probabilistic nature of the system under consideration. Successful input modeling requires a close match between the input model and the true underlying probabilistic mechanism associated with the system. The input data analysis is to model an **element** (e.g., arrival process, service times) in a discrete-event simulation given a data set collected on the element of interest. This stage performs intensive error checking on the input data, including external, policy, random, and deterministic variables. System simulation experiments aim to learn about its behavior. Careful planning, or designing, of simulation experiments is generally a great help, saving time and effort by providing efficient ways to estimate the effects of changes in the model's inputs on its outputs. Statistical experimental design methods are mostly used in the context of simulation experiments.

- **Performance evaluation and "what-if" analysis:** The "what-if" analysis involves trying to run the simulation with different input parameters (i.e., under different "scenarios") to see how performance measures are affected.

- **Sensitivity estimation:** Users must be provided with affordable techniques for sensitivity analysis if they are to understand the relationships and tradeoffs between system parameters and performance measures in a meaningful way that allows them to make good system design decisions.

- **Optimization:** Traditional optimization techniques require gradient estimation. As with sensitivity analysis, the current approach for optimization requires intensive simulation to construct an approximate surface response function. Sophisticated simulations incorporate gradient estimation techniques into convergent algorithms such as Robbins–Monroe in order to efficiently determine system parameter values which optimize performance measures. There are many other gradient estimation (sensitivity analysis) techniques, including: local information, structural properties, response surface generation, the goal-seeking problem, optimization, the "what-if" problem, and meta-modeling.

- **Report generating:** Report generation is a critical link in the communication process between the model and the end user.

Figure 1.4 shows a block diagram describing the development process for systems simulation. It describes each stage of the development as above.

1.4.1 Overview of a Modeling Project Life Cycle

A *software life cycle model* (SLCM) is a schematic of the major components of software development work and their interrelationships in a graphical framework

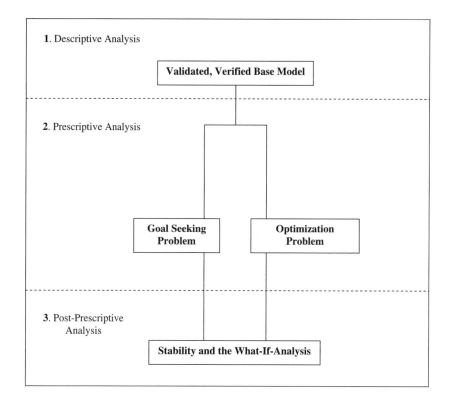

Figure 1.4 Development process for simulation

that can be easily understood and communicated. The SLCM partitions the work to be done into manageable work units. Having a defined SLCM for your project allows you to:

1. Define the work to be performed.
2. Divide up the work into manageable pieces.
3. Determine project milestones at which project performance can be evaluated.
4. Define the sequence of work units.
5. Provide a framework for definition and storage of the deliverables produced during the project.
6. Communicate your development strategy to project stakeholders.

An SLCM achieves this by:

1. Providing a simple graphical representation of the work to be performed.

2. Allowing focus on important features of the work, downplaying excessive detail.
3. Providing a standard work unit hierarchy for progressive decomposition of the work into manageable chunks.
4. Providing for changes (tailoring) at low cost.

Before specifying how this is achieved, we need to classify the life cycle processes dealing with the software development process.

1.4.2 Classifying Life Cycle Processes

Figure 1.5 shows the three main classes of a software development process and gives examples of the members of each class:

- **Project support processes:** are involved with the management of the software development exercise. They are performed throughout the life of the project.
- **Development processes:** embody all work that directly contributes to the development of the project deliverable. They are typically interdependent.
- **Integral processes:** are common processes that are performed in the context of more than one development activity. For example, the review process is performed in the context of requirements definition, design, and coding.

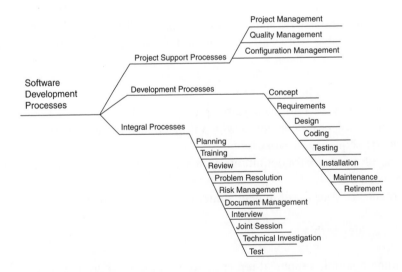

Figure 1.5   Software life cycle process classifications

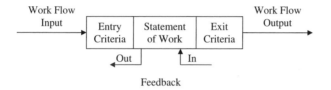

Figure 1.6 Components of a work unit

1.4.3 Describing a Process

Processes are described in terms of a series of work units. Work units are logically related chunks of work. For example, all preliminary design effort is naturally chunked together. Figure 1.6 describes the components of a work unit.

A work unit is described in terms of:

- **Work flow input/output:** Work flows are the work products that flow from one work unit to the next. For example, in Figure 1.6, the design specification flows from the output of the design work unit to the input of the code work unit. Work flows are the deliverables from the work unit. All work units must have a deliverable. The life cycle model should provide detailed descriptions of the format and content of all deliverables.
- **Entry criteria:** The conditions that must exist before a work unit can commence.
- **Statement of work (SOW):** The SOW describes the work to be performed on the work flow inputs to create the outputs.
- **Exit criteria:** The conditions that must exist for the work to be deemed complete.

The above terms are further explained through the usage of examples in Table 1.2.

Feedback paths are the paths by which work performed in one work unit impacts work either in progress or completed in a preceding work unit. For example, the model depicted in Figure 1.7 allows for the common situation where the act of coding often uncovers inconsistencies and omissions in the design. The issues raised by programmers then require a reworking of the baseline design document.

Defining feedback paths provides a mechanism for iterative development of work products. That is, it allows for the real-world fact that specifications and code are seldom complete and correct at their first approval point. The feedback path allows for planning, quantification, and control of the rework effort. Implementing feedback paths on a project requires the following procedures:

1. A procedure to raise issues or defects with baseline deliverables.
2. A procedure to review issues and approve rework of baseline deliverables.
3. Allocation of budget for rework in each project phase.

Table 1.2   Examples of entry and exit criteria and statements of work

Entry criteria	Statement of work	Exit criteria
1. Prior tasks complete	1. Produce deliverable	1. Deliverable complete
2. Prior deliverables approved and baselined	2. Interview user	2. Deliverable approved
3. Tasks defined for this work unit	3. Conduct review	3. Test passed
4. Deliverables defined for this work unit	4. Conduct test	4. Acceptance criteria satisfied
5. Resources available	5. Conduct technical investigation	5. Milestone reached
6. Responsibilities defined	6. Perform rework	
7. Procedures defined		
8. Process measurements defined		
9. Work authorized		

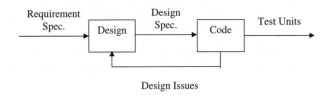

Figure 1.7   Example of a feedback path

1.4.4 Sequencing Work Units

Figure 1.8 provides an example of a life cycle model constructed by sequencing work units. We see work flows of *requirements specification* and *design specification*. The work units are design, code, and test, and the feedback paths carry requirements, design, and coding issues.

Note that the arrows in a life cycle model do not signify precedence. For example, in Figure 1.8, the design does not have to be complete for coding to commence.

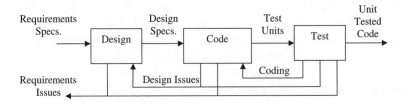

Figure 1.8   Life cycle model

Design may be continuing while some unrelated elements of the system are being coded.

1.4.5 Phases, Activities, and Tasks

A primary purpose of the life cycle model is to communicate the work to be done among human beings. Conventional wisdom dictates that, to guarantee comprehension, a single life cycle model should therefore not have more than nine work units. Clearly this would not be enough to describe even the simplest projects.

In large-scale projects, the solution is to decompose large work units into a set of levels with each level providing more detail about the level above it. Figure 1.9 depicts three levels of decomposition:

1. **The phase level:** Phases describe the highest levels of activity in the project. Examples of phases might be requirements capture and design. Phases are typically used in the description of development processes.
2. **The activity level:** Activities are logically related work units within a phase. An activity is typically worked on by a team of individuals. Examples of activities might include interviewing users as an activity within the requirements capture phase.
3. **The task level:** Tasks are components of an activity that are typically performed by one or two people. For example, conducting a purchasing manager interview is a specific task that is within the interviewing users activity. Tasks are where the work

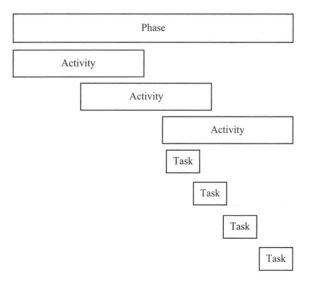

Figure 1.9   Work unit decomposition

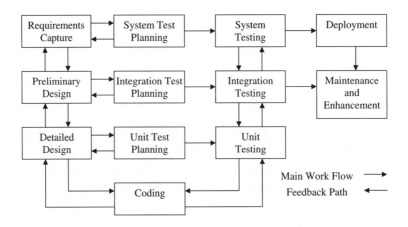

Figure 1.10   A generic project model for software development

is done. A task will have time booked to it on a time sheet. Conventional guidelines specify that a single task should be completed in an average elapsed time of 5 working days and its duration should not exceed 15 working days.

Note that phases, activities, and tasks are all types of a work unit. You can therefore describe them all in terms of entry criteria, statement of work, and exit criteria.

As a concrete example, Figure 1.10 provides a phase-level life cycle model for a typical software development project.

1.5 Summary

In this chapter, we introduced the foundations of modeling and simulation. We highlighted the importance, the needs, and the usage of simulation and modeling in practical and physical systems. We also discussed the different approaches to modeling that are used in practice and discussed at a high level the methodology that should be followed when executing a modeling project.

Recommended Reading

[1] J. Sokolowski and C. Banks (Editors), *Principles of Modeling and Simulation: A Multidisciplinary Approach*, John Wiley & Sons, Inc., 2009.

[2] J. Banks (Editor), *Handbook of Simulation: Principles, Methodology, Advances, Applications, and Practice*, John Wiley & Sons, Inc., 1998.

[3] B. Zeigler, H. Praehofer, and T.-G. Kim, *Theory of Modeling and Simulation*, 2nd Edition, Academic Press, 2000.

[4] A. Law, *Simulation Modeling and Analysis*, 4th Edition, McGraw-Hill, 2007.

2

Designing and Implementing a Discrete-Event Simulation Framework

In software engineering, the term *framework* is generally applied to a set of cooperating classes that make up a reusable foundation for a specific class of software: frameworks are increasingly important because they are the means by which systems achieve the greatest code reuse. In this section, we will put together a basic discrete-event simulation framework in Java. We call this the framework for discrete-event simulation (FDES). The framework is not large, requiring fewer than 250 lines of code, spread across just three classes, and yet, as we will see, it is an extremely powerful organizational structure. We will deduce the design of the simulation framework (together with its code) in the pages that follow. Then we will present some "toy examples" as a tutorial on how to write applications over the framework. In the next chapter, we will consider a large case study that illustrates the power of the framework in real-world settings. However, by looking at everything down to the framework internals, we will have a much better understanding of how discrete-event simulations work. In later chapters, the framework will be extended to support specialized constructs and primitives necessary for network simulation.

As noted earlier, the heart of any discrete-event simulator is a "scheduler," which acts as the centralized authority on the passage of time. The scheduler is a repository of future actions and is responsible for executing them sequentially. On the surface, this might seem like a trivial task. The complication arises from the fact that each event that is executing "now" can cause future events to be scheduled. These future events are

Network Modeling and Simulation   M. Guizani, A. Rayes, B. Khan and A. Al-Fuqaha
© 2010 John Wiley & Sons, Ltd.

typically a non-deterministic function of the present state of the entities taking part in the simulation.

2.1 The Scheduler

Let us begin by defining the `Scheduler` class:

```
public final class Scheduler implements Runnable { }
```

By making the scheduler implement the runnable interface, we will be able to assign it its own thread of execution. Typically there should be only one scheduler in any given application. We can achieve this constraint via Java language constructs using the singleton design pattern:

```
private static Scheduler _instance;
public static Scheduler instance() {
  if (_instance == null) { _instance = new Scheduler(); }
  return _instance;
}
private Scheduler() {}
```

Now the only way to access a scheduler is via the static method `Scheduler.instance()`, which makes a scheduler instance if one has not already been made (or returns the existing one if one has been made previously)—no external class is permitted to make a scheduler since the constructor is private.

A scheduler's notion of time is punctuated by events, and events are defined to be interactions between two entities at a particular time—one of which is considered the initiator and the other is denoted the target. We note that this one-to-one model does not cover all types of influence. In particular, one-to-many and many-to-many influences are not covered directly. However, initially, these more complex forms of interaction can be simulated using higher-level constructs built "on top" of the scheduler we will put together.

The entities in our simulation will be represented in our code by a base abstract Java class called `SimEnt`. The code for this `SimEnt` class will be specified later in this section, and the reader need only keep this in mind right now. The influence transmitted between one (initiator) `SimEnt` and another (target) `SimEnt` will be embodied in classes implementing the event interface. The event interface will also be specified later.

To begin, we need to make concrete the scheduler's notion of an event, which needs to be something more than just the event (the influence). The reason for this is that the scheduler acts as a mediator between the initiator and the target, and the mediator is responsible for accepting the event from the former and delivering it to the latter at the appropriate time. To this end, we define a class called EventHandle, which holds all this information, in addition to the event itself:

```
public static class EventHandle {
  private final SimEnt _registrar, _target;
  private final Event _ev;
  private final UniqueDouble _udt;

  private EventHandle(SimEnt registrar, SimEnt target,
    Event ev, UniqueDouble udt) {
    _registrar = registrar;
    _target = target;
    _ev = ev;
    _udt = udt;
  }
}
```

Since this is internal to the scheduler, it is best to make it a static nested inner class. The _udt field is meant to hold the time at which the event _ev will be delivered by the scheduler to the _target _siment. The reader may wonder what a UniqueDouble type is, since it is not part of the standard Java built-in classes. The answer to this lies in the fact that even though two events may be scheduled for the same time, in actual implementation they should happen in causal order. More precisely, if SimEnt A schedules an event E1 to SimEnt B for time 10, and then SimEnt C schedules an event E2 to SimEnt B for time 10 also, the delivery of E1 must occur before the delivery of E2, even though they occur at the same clock time. To achieve this requires making an extended class which holds both time and a discriminating field which can be used to determine the relative orders of E1 and E2. Below is one way to achieve this:

```
// registrations for the same time.
private int _uid=0;

private static class UniqueDouble implements Comparable {
  Double _value;
  int _discriminator;

UniqueDouble(double value) {
  _value = new Double(value);
  _discriminator = (Scheduler.instance()._uid);
```

```
    (Scheduler.instance()._uid)++;
}

public int compareTo(Object obj) {
  UniqueDouble other = (UniqueDouble)obj;
  // compare by value
  if (this._value.doubleValue() < other._value.doubleValue
    ())
    return -1;
  else if (this._value.doubleValue() >
      other._value.doubleValue())
    return +1;
  else { // but use the discriminator if the values agree
    if (this._discriminator < other._discriminator)
      return -1;
    else if (this._discriminator > other._discriminator)
      return +1;
    else
      return 0;
  }
}

public String toString() {
  return _value.toString()+"("+_discriminator+")";
  }
}
```

Again, since this notion of time is internal to the scheduler, it is best to implement UniqueDouble as a static nested inner class inside the scheduler.

We are now ready to implement the mechanism by which the scheduler manages its EventHandle objects corresponding to pending actions. It will be helpful to organize these EventHandle objects in three distinct ways, for efficiency: (1) by the initiator SimEnt; (2) by the target SimEnt; and (3) by the time at which the delivery is to take place. The first two of these can each be implemented as a HashMap from SimEnt to a HashSet of EventHandle objects:

```
private final HashMap _from2set = new HashMap();
private Set getEventsFrom(SimEnt e) {
  HashSet set = (HashSet)_from2set.get(e);
  if (set == null) {
    set = new HashSet();
```

```
    _from2set.put(e, set);
  }
  return set;
}

private final HashMap _to2set = new HashMap();
private Set getEventsTo(SimEnt e) {
  HashSet set = (HashSet)_to2set.get(e);
  if (set == null) {
    set = new HashSet();
    _to2set.put(e, set);
  }
  return set;
}
```

The third (organizing `EventHandle` objects by delivery time) can be using a `TreeMap` from delivery time to `EventHandle`:

```
// UniqueDouble(time)->Event
private final TreeMap _ud2ehandle = new TreeMap();
```

With these data structures and supporting methods in place, we are now ready to define methods for registering an event delivery. The following method takes as arguments the initiator (registrar), the target, the event `ev`, and the time (from now) when `ev` should be delivered:

```
public static Scheduler.EventHandle register (SimEnt
  registrar,
    _SimEnt target,
    Event ev, double t) {
  double deliveryTime = Scheduler.getTime() + t;
  EventHandle handle = new EventHandle(registrar, target, ev,
    new UniqueDouble(deliveryTime));
  instance().getEventsFrom(handle._registrar).add(handle);
  instance().getEventsTo(handle._target).add(handle);
  instance()._ud2ehandle.put(handle._udt, handle);
  return handle;
}
```

The method returns the `EventHandle` that is used internally in the scheduler's maps. Note that using event handles as the key/values of the scheduler's maps also allows us to use the same event object within distinct registrations (at different times). Returning an `EventHandle` allows the caller of `register()` to later "deregister" the event, thereby canceling its delivery:

```
public static void deregister(EventHandle handle) {
  instance().getEventsFrom(handle._registrar).remove
    (handle);
  instance().getEventsTo(handle._target).remove(handle);
  instance()._ud2ehandle.remove(handle._udt);
}
```

Let us now implement the main loop of the scheduler:

```
private boolean _done = false;
private double _timeNow = 0;

  public void run() {
    do {
      if (_ud2ehandle.size() == 0) _done = true;
      else {
        UniqueDouble udt = (UniqueDouble)_ud2ehandle.
          firstKey();
        EventHandle h = (EventHandle)_ud2ehandle.get(udt);
        _timeNow = udt._value.doubleValue();
        h._ev.entering(h._target);
        h._target.recv( h._registrar, h._ev );
        h._registrar.deliveryAck( h );
        deregister(h);
      }
    }
  while (!_done);
  killall();
  reset();
}
```

The main loop simply runs through the `_ud2ehandle` map, removing the next `EventHandle`, advancing the time. It notifies the event of its impending delivery to the target using the entering `method()`. It also notifies the target `SimEnt` of the event's arrival via its `recv()` method. Finally, it notifies the initiator `SimEnt` of the event's delivery by calling its `deliverAck()` method. Once there are no more events to be delivered (and push simulation time forward), the `_done` variable is set to false and the loop ends. Once outside the loop, the `SimEnts` are notified that the simulation is over via the `killall()` method:

```
private void killall() {
  while (_from2set.keySet().size() > 0) {
    SimEnt se = null;
    Iterator it=_from2set.keySet().iterator();
    se = (SimEnt)it.next();
      se.suicide();
  }
}
```

The scheduler then clears its data structures using the `reset()` method, and the `run()` method terminates, causing its thread of execution to complete. The scheduler maintains its notion of the present time in the `_timeNow` variable. This can be revealed to outside classes via a public getter method:

```
public double getTime() { return _timeNow; }
```

The main event loop can be stopped externally by calling the `stop()` method.

```
public void stop() { _done = true; }
```

To be complete, we need to consider the possibilities that not all `SimEnt` instances may exist at the outset, since they may need to be created dynamically in the course of the simulation itself (e.g., in an ecological simulation, children are created as time passes). Moreover, some `SimEnt` instances may be destroyed in the course of the simulation (e.g., in an ecological simulation, some individuals die as time passes). What should the scheduler do when a `SimEnt` is "born" or "dies?" Certainly, any events that are pending for a dead `SimEnt` should be removed from the Scheduler:

```
void deathSimEnt(SimEnt ent) {
  // clone to avoid concurrent modifications from deregister
  Set from = new HashSet(getEventsFrom(ent));
  for (Iterator it = from.iterator(); it.hasNext();) {
    EventHandle h = (EventHandle)it.next();
    deregister(h);
  }
  _from2set.remove(ent);

  // clone to avoid concurrent modifications from deregister
```

```
Set to = new HashSet(getEventsTo(ent));
for (Iterator it = to.iterator(); it.hasNext();) {
  EventHandle h = (EventHandle)it.next();
  deregister(h);
}
_to2set.remove(ent);
}
```

What about the case when a new SimEnt is created? In this case, we do not really need to do anything, but let us put empty sets for the new SimEnt in the _from2set and _to2set maps by calling getEventsFrom() and getEventsTo():

```
void birthSimEnt(SimEnt ent) {
  // make the sets by getting them
  Set from = instance().getEventsFrom(ent);
  Set to = instance().getEventsTo(ent);
}
```

Because of the above code, we can be certain that if a SimEnt exists then it appears as a key in the _from2set and _to2set maps. Finally, it may occasionally be necessary to reset the scheduler to its initial state. This is now easy, since all that is needed is to simulate the death of all SimEnt instances, zero out the maps, and reset the variables to zero:

```
public void reset() {
  this._from2set.clear();
  this._to2set.clear();
  this._ud2ehandle.clear();
  this._timeNow = 0;
  this._uid = 0;
  this._done = false;
}
```

This completes the exposition of the scheduler's class. We now go on to exposit the base abstract SimEnt class from which all concrete SimEnt classes are to be derived.

2.2 The Simulation Entities

We now describe the base SimEnt class. A SimEnt represents a unit of logic which responds to the arrival of events. The response can involve: (1) sending events,

(2) creating new `SimEnts`, and (3) destroying `SimEnts`. A `SimEnt` which sends an event will receive a confirmation or "Ack" from the scheduler once the event has been delivered. We begin with an empty class and gradually fill in its definition:

```
public abstract class SimEnt {
}
```

The construction of `SimEnt` requires registration with the scheduler, and its death requires deregistration with the scheduler. This is ensured by providing the following protected constructor and `suicide()` method:

```
protected SimEnt() {
  Scheduler.instance().birthSimEnt(this);
}

protected final void suicide() {
  this.destructor();
  Scheduler.instance().deathSimEnt(this);

}
  protected void destructor() {
    // default no op
}
```

Although the `suicide()` method is final, derived classes can define their own `destructor()` function which will get called whenever the `SimEnt` is "suicided." Derived `SimEnt` classes will need to be able to schedule and cancel the delivery of events. This facility is provided via two protected methods:

```
protected final Scheduler.EventHandle
  send(SimEnt dst, Event ev, double t) {
    return Scheduler.instance().register(this, dst, ev, t);
  }

protected final void revokeSend(Scheduler.EventHandle h) {
  Scheduler.instance().deregister(h);
}
```

Finally, derived `SimEnt` classes must respond to the arrival of events (sent by other `SimEnts`). They will also receive acknowledgements of the delivery of events that they have sent. These two responses are specified in derived `SimEnts` by providing a definition of the following `public abstract` methods:

```
public abstract void recv(SimEnt src, Event ev);
public abstract void deliveryAck(Scheduler.EventHandle h);
```

The above two methods are called by the scheduler inside its `run()` loop, where the target's `recv()` is called and the initiator's `deliveryAck()` is called. This completes the specification of the base abstract `SimEnt` class. To be a concrete subclass of `SimEnt`, all that is needed is to implement the above two abstract methods, which provide a specification of how the `SimEnt` responds to receiving arrival and confirmation of its own send requests.

2.3 The Events

Events are the objects which embody the influence of one `SimEnt` on another `SimEnt`. Their timed delivery is mediated by the scheduler. The scheduler needs events to support one method:

```
public interface Event {
  public void entering(SimEnt locale);
};
```

The scheduler calls this method to allows an event to make last-minute adjustments to itself just prior to being delivered to the target. This circumvents the problem where the initiator of the event's delivery may not have been able to fully specify it at the time when the delivery was scheduled. (For example, suppose a `SimEnt` wants to schedule a message now that will be delivered to another `SimEnt` tomorrow, and whose purpose is to withdraw all the money in a bank account. The initiator cannot fully craft this message now, since the amount of the withdrawal will not be known until tomorrow.)

2.4 Tutorial 1: Hello World

To illustrate how the previous discrete-event simulation framework can be used, let us write the equivalent of "Hello World" for it. Below is a simple `SimEnt` called `Agent`, which upon creation registers a `HelloWorld` event to be sent to itself in 10 seconds:

```
public class Agent extends SimEnt {
  Agent() {
    super();
    send(this, new HelloWorld(), 10.0);
```

```
  }
  public void recv(SimEnt src, Event ev) {
    System.out.println("Hello World message received.");
  }

  public String getName() {
    return "Agent";
  }

  public void deliveryAck(Scheduler.EventHandle h) {
    System.out.println("Hello World delivery complete.");
  }
}
```

Notice that `Agent` is a concrete instance of `SimEnt`, defining the two abstract methods of its parent class: `recv()`, `getName()`, `recv()`, and `deliveryAck()`. In its constructor, `Agent` sends itself a new `HelloWorld` event, scheduled for 10 simulation seconds from now. The `send()` method is of course inherited from the base `SimEnt` class. Let us specify the `HelloWorld` event class now:

```
public class HelloWorld implements Event {
  public void entering(SimEnt locale) {
    double t = Scheduler.getTime();
    System.out.println("Hello World entering target at
      time: "+t);
  }
}
```

And let us see how to put it all together in a main class:

```
public class Run {
  public static void main(String [ ] args) {
    Agent a = new Agent();

    Thread t = new Thread(Scheduler.instance());
    t.start();
    try { t.join(); }
    catch (Exception e) { }
  }
}
```

We compile the following code:

```
javac -g -d ./classes -classpath ../../classes/fw.jar *.java
```

and when we execute the code we see:

```
java -cp ./classes:../../classes/fw.jar fw.ex1.Run
```

```
Hello World entering target at time: 10
Hello World message received.
Hello World delivery complete.
```

2.5 Tutorial 2: Two-Node Hello Protocol

The next example builds in complexity, by introducing two SimEnt instances (of class Node), which send each other messages at regular intervals. The way this is achieved is to have each node send itself a clock event periodically, and whenever this clock event is received, the node sends its peer node a HelloMsg event.

Below is the code of the Node class. Notice that the Node sends itself a Clock in its constructor. In addition, in the recv() method, if it gets a Clock event, it sends its peer a HelloMsg, and then sends itself the same Clock event (ev) for delivery in 10 seconds. Once it has sent eight HelloMsg events to its peer, it calls suicide():

```
public class Node extends SimEnt {

  private int _id;
  public Node(int id) {
    super();
    _id = id;
    send(this, new Clock(), 10.0);
  }

  private SimEnt _peer;

  public void setPeer(SimEnt peer) {
    _peer = peer;
  }

  private int _sentmsg=0;
  private int _recvmsg=0;
```

```
public void recv(SimEnt src, Event ev) {
  if (ev instanceof Clock) {
    send(_peer, new HelloMsg(), 10.0);
    _sentmsg++;
    if (_sentmsg<=8) send(this, ev, 10.0);
    else this.suicide();
  }
  if (ev instanceof HelloMsg) {
    _recvmsg++;
    System.out.println("Node "+_id+" recv msg "+_recvmsg);
  }
}

public String getName() {
  return "Node";
}

public void deliveryAck(Scheduler.EventHandle h) { }
}
```

Although Node will receive an acknowledgement of the delivery of the HelloMsg events through its deliveryAck() method, for simplicity we do not make it respond to this notification.

Below is the HelloMsg event that is passed between Node instances:

```
public class HelloMsg implements Event {
  public void entering(SimEnt locale) {}
}
```

Again, event though HelloMsg instances will be notified by the scheduler just prior to being delivered to a SimEnt via the entering() method, for simplicity we do not make our HelloMsg react to these notification. Let us put the whole system together now. The process is quite simple. We make two nodes, each informing the other, then we start the scheduler:

```
public class Run {

  public static void main(String [ ] args) {
    Node n1 = new Node(1);
    Node n2 = new Node(2);
```

```
    n1.setPeer(n2);
    n2.setPeer(n1);

    Thread t = new Thread(Scheduler.instance());
    t.start();
    try { t.join(); }
    catch (Exception e) { }
  }
}
```

Compiling is achieved via:

```
javac -g -d ./classes -classpath ../../classes/fw.jar *.java
```

Execution yields the following output:

```
java -cp ./classes:../../classes/fw.jar fw.ex2.Run
```

```
Node 2 recv msg 1
Node 1 recv msg 1
Node 2 recv msg 2
Node 1 recv msg 2
Node 2 recv msg 3
Node 1 recv msg 3
Node 2 recv msg 4
Node 1 recv msg 4
Node 2 recv msg 5
Node 1 recv msg 5
Node 2 recv msg 6
Node 1 recv msg 6
Node 2 recv msg 7
Node 1 recv msg 7
Node 2 recv msg 8
```

2.6 Tutorial 3: Two-Node Hello through a Link

This next example builds on the last one by having both nodes send their HelloMsg events to an intermediate Link object which connects the two. The Link must of course be a SimEnt as well, in order to be able to receive and send the HelloMsg events. Below is the code for the `Link`:

```
public class Link extends SimEnt {

  public Link() {
    super();
  }

  private Node _a;
  public void setA(Node peer) {
    _a = peer;
  }

  private Node _b;
  public void setB(Node peer) {
    _b = peer;
  }

  public void recv(SimEnt src, Event ev) {
    if (ev instanceof HelloMsg) {
      System.out.println("Link recv msg, passes it
        through...");
      if (src==_a) {
        send(_b, ev, 10.0);
      }
      else {
        send(_a, ev, 10.0);
      }
    }
  }

  public String getName() {
    return "Link";
  }

  public void deliveryAck(Scheduler.EventHandle h) { }
}
```

Notice how there are two methods, `setA()` and `setB()`, which will be used for informing the `Link` about the `Node` objects at its two ends. In the `recv()` method, the Link sees if the event ev is of type HelloMsg. If so, it sends it onward to the other peer (in 10 seconds). Let us see how we modify the Run class to hook up Link to the two nodes:

```
public class Run {
  public static void main(String [ ] args) {
    Link link = new Link();

    Node n1 = new Node(1);
    n1.setPeer(link);
    link.setA(n1);
    Node n2 = new Node(2);
    n2.setPeer(link);
    link.setB(n2);

    Thread t = new Thread(Scheduler.instance());
    t.start();
    try { t.join(); }
    catch (Exception e) { }
  }
}
```

We make the two Node objects, and make the Link. The two Node objects are told that the Link is their Peer. The Link is informed about the two nodes. Then the scheduler is started. Let us see what happens. We compile:

```
javac -g -d ./classes -classpath ../../classes/fw.jar:../
  ex2/classes *.java
```

and then we run the program:

```
java -cp ./classes:../../classes/fw.jar:../ex2/classes fw.
  ex3.Run
```

```
Link recv msg, passes it through...
Link recv msg, passes it through...
Node 2 recv msg 1
Link recv msg, passes it through...
Node 1 recv msg 1
Link recv msg, passes it through...
Node 2 recv msg 2
```

```
Link recv msg, passes it through...
Node 1 recv msg 2
Link recv msg, passes it through...
Node 2 recv msg 3
Link recv msg, passes it through...
Node 1 recv msg 3
Link recv msg, passes it through...
Node 2 recv msg 4
Link recv msg, passes it through...
Node 1 recv msg 4
Link recv msg, passes it through...
Node 2 recv msg 5
Link recv msg, passes it through...
Node 1 recv msg 5
Link recv msg, passes it through...
Node 2 recv msg 6
Link recv msg, passes it through...
Node 1 recv msg 6
Link recv msg, passes it through...
Node 2 recv msg 7
Link recv msg, passes it through...
Link recv msg, passes it through...
Node 1 recv msg 7
```

2.7 Tutorial 4: Two-Node Hello through a Lossy Link

The final example extends the previous one by making a more specialized kind of link—one that sometimes drops messages. The definition of such a LossyLink class requires specifying the probability with which an event will be dropped. By making LossyLink a subclass of Link, we can reuse the code from the Link class. Below is the code for LossyLink. Notice that it takes the loss probability in its constructor, and uses an object of type Random() to determine whether a received event should be dropped, or delegated to its parent's implementation (in the Link class):

```
public class LossyLink extends Link {
    private final double _lossp;
    private final Random _random;

    public LossyLink(double lossp) {
      super();
```

```
    _lossp = lossp;
    _random = new Random();
  }

  public void destructor() {
    System.out.println("LossyLink dropped "+_dropped+"
      packets total.");
  }
  private int _dropped = 0;
  public void recv(SimEnt src, Event ev) {
    if (_random.nextDouble() < _lossp) {
    _dropped++;
    System.out.println("Link drops msg...");
  }
  else super.recv(src,ev);
}

public String getName() {
  return "LossyLink";
}
}
```

We have added a destructor to the class to illustrate when it gets called. Now, making a LossyLink network of two nodes is analogous to when there was a Link:

```
public class Run {

  public static void main(String [ ] args) {
    Link link = new LossyLink(0.3);

    Node n1 = new Node(1);
    n1.setPeer(link);
    link.setA(n1);

    Node n2 = new Node(2);
    n2.setPeer(link);
    link.setB(n2);

    Thread t = new Thread(Scheduler.instance());
    t.start();
```

```
    try { t.join(); }
    catch (Exception e) { }
  }
}
```

Compiling the code via:

```
javac -g -d ./classes -classpath
../../classes/fw.jar:../ex2/classes:../ex3/classes *.
  java
```

and then executing it, we have:

```
java -cp ./classes:../../classes/fw.jar:../ex2/
classes:../ex3/classes fw.ex4.Run

Link recv msg, passes it through...
Link drops msg...
Node 2 recv msg 1
Link drops msg...
Link recv msg, passes it through...
Link recv msg, passes it through...
Node 1 recv msg 1
Link recv msg, passes it through...
Node 2 recv msg 2
Link recv msg, passes it through...
Node 1 recv msg 2
Link drops msg...
Node 2 recv msg 3
Link recv msg, passes it through...
Link drops msg...
Node 2 recv msg 4
Link drops msg...
Link recv msg, passes it through...
Link recv msg, passes it through...
Node 1 recv msg 3
Link recv msg, passes it through...
Node 2 recv msg 5
Link recv msg, passes it through...
Link drops msg...
LossyLink dropped 6 packets total.
```

2.8 Summary

In this chapter, we have designed three cooperating classes that together serve as a reusable foundation for a wide range of simulation software to be developed in subsequent chapters. The three classes were: the `Scheduler`, which manages the discrete passage of time; the `SimEnt`, which is the base abstract entity within the simulation; and the `Event`, which is a temporally defined occurrence at a `SimEnt`. The framework we developed will be referred to in future chapters as the framework for discrete-event simulation (FDES). The FDES is quite compact, requiring fewer than 250 lines of code, but powerful in that it provides an organizational structure for discrete simulations. In the next chapter, we will consider a large case study that illustrates the power of the framework in real-world settings. In later chapters, the framework will be extended to support specialized constructs and primitives necessary for network simulation.

Recommended Reading

[1] E. Gamma, R. Helm, R. Johnson, and J. Vlissides, *Design Patterns: Elements of Object-Oriented Software*, Addison-Wesley, 1995.

[2] L. P. Deutsch, "Design reuse and frameworks in the Smalltalk-80 system," in *Software Reusability, Volume II: Applications and Experience* (T. J. Biggerstaff and A. J. Perlis,eds.), Addison-Wesley, 1989.

[3] R. E. Johnson and B. Foote, "Designing reusable classes," *Journal of Object-Oriented Programming*, vol. 1, pp. 22–35, June/July 1988.

3

Honeypot Communities: A Case Study with the Discrete-Event Simulation Framework

In this chapter, we will design and develop a simulation of a system that will generate malware antivirus signatures using an untrusted multi-domain community of honeypots. Roughly speaking, the honeypot community will act as a "Petri dish" for worms. The distribution of sensors within the Internet will ensure that this Petri dish is a microcosm reflecting overall worm demographics. On the other hand, the small size of the Petri dish will accelerate infection percolation rates, allowing automated antidote generation to become feasible well in advance of when the worm disables the ambient larger network.

The system described here is based on an earlier system that was developed by the authors and described in [1]. Unlike that publication, however, which focused on experimental results and their implications, here the exposition is simplified and entirely focused on the design and implementation of the necessary simulation software itself. The purpose is to provide the reader with a real case study illustrating how builds a simulation using the Discrete Event Simulation Framework that we designed and implemented in Chapter 2. Before we develop the simulation of the proposed honeypot-based system, let us begin with some background, so that the motivation of the problem to be solved is clear, and the design of the proposed solution is compelling.

3.1 Background

Since the 1990s, computer worms have attacked the electronic community with alarming regularity. Prominent examples include Melissa (1999), Code Red (2001), Slammer (2003), Blaster (2003), and Sasser (2004), to name just a few! While the economic impact of these worm attacks is believed to be in excess of billions of dollars,

Network Modeling and Simulation   M. Guizani, A. Rayes, B. Khan and A. Al-Fuqaha
© 2010 John Wiley & Sons, Ltd.

there is no sure antidote against newly emergent worms. As such, society as a whole remains vulnerable to the dangers that malware poses. Network worms spread over the Internet by accessing services with exploitable implementation flaws or "vulnerabilities." Newly infected hosts serve as a stepping stone, advancing the infection and leading to thousands of vulnerable hosts becoming compromised in a very short time. Since worms have historically been static in their propagation strategies, an attack against a vulnerable host typically follows a predictable pattern or *signature*. Modem intrusion detection systems (IDSs) use the latter, by matching the ports and byte sequences of incoming traffic to a specific signature in order to identify worm traffic as it arrives and prevent the vulnerable services from seeing virulent packets. Examples of popular IDSs include Bro [2] and Snort.

While signatures-based IDSs are successful against threats that are known in advance, they remain useless against new worms because well-designed network worms can propagate much more quickly than signatures can be generated. Most commercial solutions rely on hand-generated signatures, which is a labor-intensive process that takes on the order of hours or days. In contrast, modem worms spread fast (e.g., the Slammer [3] worm can double its number of infections every 8.5 seconds and can infect more than 90% of vulnerable hosts within 10 minutes). A comprehensive solution to the worm problem must be able to detect new worms and rapidly provide an active response without labor-intensive human interaction.

A number of approaches for detecting and responding to worms have been considered. The idea of collaborative intrusion detection itself is not new. The Worminator system uses alerts that are shared within a distributed IDS in order to detect an attack in progress. The DOMINO system makes a strong case for collaborative intrusion detection by demonstrating that blacklists can be constructed more quickly using a large, distributed, collaborative system. However, the veracity of shared information is difficult to verify in such schemes; as is usually the case, participants are required to trust one another.

Detection systems use a passive network telescope [5]—blocks of previously unallocated IP addresses—to capture scan traffic. Such detection allows only the most crude form of response since all clients must be blocked from access to a vulnerable service until it has been secured. Collaborative detectors of this form also require total trust among all participants.

Content filtering systems attempt to stop infections at the *destination*. Signature systems such as EarlyBird and Autograph [7] observe flows from infected machines, identify blocks common across many of those flows, and dynamically create IDS signatures to enable content filtering. In order to generate signatures, a source of malicious flows is required. Autograph requires a large deployment to obtain malicious flows rapidly. The Honeycomb [8] project collects malicious flows through medium-interaction honeypots. Both systems require total trust among participants.

3.2 System Architecture

There are many challenges in developing a scalable and comprehensive solution that is capable of thwarting worm propagation:

- The system must be able to detect worms early, before they become widespread.
- The system must be accurate: there should be no false negatives (no worms should go undetected) and no false positives (since network messages misdiagnosed as worms can cause denied service).
- The system should automatically generate countermeasures that can be fully tested and guaranteed to be effective in catching not only known worms, but also new worms that may be engineered to evade or subvert the system.
- The system must be practical, reliable, and easily deployed and maintained.

In addition to these challenges, a serious obstacle is adoption. Network worms are fast enough to overrun any detector deployed in a single administrative domain, thus making the case for a distributed, collaborative detector. Collaboration that requires a high degree of trust among participants is unlikely to be deployed, hence our goal is:

Maximize coverage in the network, while minimizing trust requirements.

Our system works as follows: a set of sensor machines and honeypot machines are deployed throughout the wide area. Sensors divert all unsolicited traffic by tunneling it to one or more honeypots. If a worm successfully infects a honeypot, it will attempt to make outgoing infection attempts to propagate itself. The host honeypot, however, tunnels all outgoing attacks *toward other honeypots*. Once many honeypots have been infected, a signature can be developed, and the people administering these honeypots can learn of the platform and services affected by the worm and the signatures developed. Our system is designed to be transparently distributed across multiple networks, allowing participants everywhere to join. A machine can join the system by becoming a sensor or a honeypot. In contrast to prior efforts, participants in this system share actual infections (rather than just sharing information) which can be locally verified using honeypots. The system maximizes participation by minimizing trust required among the participants.

We now describe the components and functions of the system in more detail.

Sensors are machines configured to send "Unwanted" traffic into a honeypot. Unwanted traffic is taken to be any traffic that would otherwise be dropped by the sensor. Examples include unsolicited connections (TCP SYN packets), UDP packets to unused ports, incoming ICMP requests, and so on, all of which are sent to a honeypot by means of an SSL tunnel. Sensors are cheap since existing machines and firewalls at the front of the network can be configured to route unwanted traffic for many IP addresses to distinct honeypots.

Honeypots are machines with known vulnerabilities deployed in the network to learn about attackers' motivations and techniques. In our system, participants are free to deploy honeypots in any configuration, but for the detection of new worms we expect that honeypots will be up to date on patches and thus only exploitable via previously unknown vulnerabilities. Specifically, participants will deploy honeypots for services they are interested in protecting. Each honeypot needs an associated manager process that is responsible for communicating with other hosts via the overlay network, setting up tunnels with sensors and other managers, recording outbound traffic, and generating signatures when infection occurs. If the honeypot is implemented using virtualization technology like VMWare or Xen [9], then it can reside on the same physical machine as the manager. Though deploying honeypots is more expensive than deploying sensors, honeypots allow the participants to observe infections on a machine they control and thus obtain conclusive proof of a threat.

When a sensor/honeypot joins the system, it builds tunnels to honeypots by contacting its manager and negotiating the details of the tunnel interface (e.g., private IP addresses, routes, IP address of the honeypot, traffic policies, encryption). We assume honeypots are heterogeneous in this case. Our system would use a "Chord distributed hash table lookup" primitive [10] for sensors/honeypots to locate honeypots on a particular platform or hosting a particular service.

By manipulating IP table rules, we apply destination network address translation (DNAT) to unwanted traffic for the sensor toward the address of the honeypot, forwarding the packet over the tunnel to the honeypot's manager. Many current worms (e.g., Blaster and Sasser) require the infected host to make fresh TCP connections to the infector to retrieve the executable that comprises the bulk of the worm's logic. By configuring the sensor like a masquerading firewall for the honeypot, the sensor can proxy connections from the honeypot and allow it to communicate with the infector. It is obvious to the honeypot manager when a honeypot has been infected because the largely inactive honeypot suddenly begins making many outgoing connections to random IP addresses. To spread the infection, within the "Petri culture," these infection attempts are rerouted to other honeypots. Each outbound infection attempt (originally destined for a random IP address) is "DNATed" instead to a randomly chosen honeypot over a tunnel established by the corresponding honeypot managers. The target honeypot manager acts as a masquerading firewall for the infector, and sources NAT traffic from itself so that it appears to be coming from target.

Once the infection has spread among the honeypots, each associated manager has undeniable, locally observable evidence of a threat. This information is easy to trust because it has been observed using resources that the manager controls. In addition, the honeypot manager has observed many details about the infection, including the content of any traffic the worm might generate on the network. The particular choice of active response mechanism is independent of the system architecture. In our system, a honeypot manager queries a (small) random set of peer honeypots, retrieves their observed flows, and uses these to attempt generation of a signature. Even if some participants are malicious, they cannot influence signature generation unless they are in a majority. Once a signature has been generated, its effectiveness can be tested using the live copy of the worm that has been captured. This minimizes trust requirements for generated signatures. To deploy the signature, the honeypot manager simply needs to augment the IDSs it controls.

3.3 Simulation Modeling

We now present a discrete-time event simulator to measure the efficacy of the *culturing* approach to worm detection under various parameters. The simulation considers a network consisting of S sensors, H honeypots, and O ordinary machines. Each sensor (resp. honeypot) knows only a small set of FEED_HP (PEER_HP) honeypots. Each of these $S + H + O$ machines is "vulnerable" to the worm with a probability PRM_PROB_VULNERABLE. Initially, a set of INITIAL POP worms is instantiated on distinct randomly chosen machines. Every SCAN PERIOD seconds, each worm hurls itself toward an IP address which is taken to be in the set of sensor/honeypot/ordinary machines with probability $(S + H + O)/2^{32}$. If the source of the worm is a honeypot or sensor, then tunneling is simulated by taking the destination to be a randomly chosen honeypot. If the source of the worm is an ordinary machine, then the destination is simply taken to be a random machine. Finally, if the destination of the transmission happens to be both vulnerable and uninfected, a new copy of the worm is instantiated there. While the parameters are adjustable in our simulation environment, we fix them at realistic values for the experiment.

Let us begin by making a list of the various SimEnt classes:

- Machine (an infectable computer or network element)
- Honeypot (a special kind of Machine)
- Sensor (another special kind of Machine)
- Env (a network of Machines)
- Worm (the malicious software)
- Experimenter (the nexus of statistical information from the simulation).

The most crucial element in a SimEnt's behavior is how it responds to incoming events in its recv() method. In what follows, we consider this method for each of the above SimEnts.

3.3.1 Event Response in a Machine, Honeypot, and Sensors

A machine responds to the receipt of infection events by determining if the infection is successful and, if so, becoming infected. Determination of vulnerability is calculated stochastically by a random generator. The probability that the worm succeeds in infecting the machine is given by the parameter `Machine.PRM_PROB_VULNERABLE`. Below we see the code for `recv()` in the `Machine` class:

```
// how a machine responds to events
public void recv(SimEnt src, Event ev) {
  // dispatch!
  if (ev instanceof Kill) {
    this.suicide();
  }
  else if (ev instanceof Infection) {
    if (Chance.occurs(Machine_PRM_PROB_VULNERABLE)) {
      Infection inf = (Infection)ev;
      this.infect();
    }
  }
}
```

It may also be necessary to kill a machine (e.g., at the end of the simulation), so we make the `recv()` method sensitive to the arrival of `Kill` events. What happens when the machine determines that it is infected? The `infect()` method simply instantiates a new worm that resides "in" the machine. Most real worms do not make multiple instances of themselves, so the machine checks to make sure that its worm occupancy does not exceed the parameter `Worm.PRM_MAX_INSTANCES`:

```
public void infect() {
  // we have received an infection
  if (!_infected) {
    _infected = true;
  }
  // we can take on more worms
  if (_worms.size() < Worm.PRM_MAX_INSTANCES) {
    _worms.add(new Worm(this));
    if (Machine.DIAG) {
      System.out.println("Worm arrived on "+this.name());
    }
  }
}
```

3.3.2 Event Response in a Worm

A worm is always made by a machine when it receives an `Infection` event (and probabilistically determines that the infection was successful). At the time when the worm is instantiated, it is told the identity of its host machine (or "residence"):

```
public Worm(Machine residence) {
  super();
  _residence = residence;
  TimeTick tt = new TimeTick(PRM_WORM_WAKEUP);
  this.send(this, tt, tt.getPeriod());
}
```

A worm responds to `TimeTick` events by asking its host (machine, sensor, or honeypot) for a random neighboring machine ("peer"). The worm then hurls itself at this random neighbor using an `Infection` event. As above, it may be necessary to kill a worm (e.g., at the end of the simulation), so we make the `recv()` method sensitive to the arrival of `Kill` events:

```
public void recv(SimEnt src, Event ev) {
  // dispatch!
  if (ev instanceof Kill) {
    this.suicide();
  }
  else if (ev instanceof TimeTick) {
    TimeTick tt = (TimeTick)ev;
    Machine h = _residence.getRandomPeer();
    if (h!=null) {
      // diag
      Infection inf = new Infection();
      this.send(h, inf, 0.0);
      _residence.outboundTrafficNotification();
    }
    // register next tick
    this.send(this, tt, tt.getPeriod());
  }
}
```

Note that the `getRandomPeer()` method that the worm uses to obtain the address of its next victim is overloaded in the `Sensor` and `Honeypot` subclasses of `Machine`. In this way, a machine provides a random network neighbor, while sensors always tunnel infections into honeypots, and honeypots always tunnel the infection into other honeypots.

Here is `getRandomPeer()` in `Machine`:

```
public Machine getRandomPeer() {
  if (_peerMachines.size() == 0) return null;
  int index = (int)(Math.random()*(double)_peerMachines.
    size());
  return (Machine)_peerMachines.get(index);
}
```

Compare this with `getRandomPeer()` in `Sensor`:

```
public Machine getRandomPeer() {
  // only allow outbound connections to Honeypots
  if (_feedHoneypots.size() == 0) return null;
  int index = (int)(Math.random()*(double)_feedHoneypots.
    size());
  return (Honeypot)_feedHoneypots.get(index);
}
```

and `getRandomPeer()` in `Honeypot`:

```
public Machine getRandomPeer() {
  // only allow outbound connections to Honeypots
  if (_peerHoneypots.size() == 0) return null;
  int index = (int)(Math.random()*(double)_peerHoneypots.
    size());
  return (Honeypot)_peerHoneypots.get(index);
}
```

The worm also calls `outboundTrafficNotification()` on its residence to allow the host to respond to its outward propagation attempt. Regular machines and sensors do not respond in any way to this notification (their implementation of the method is empty), but honeypots do. They attempt to construct a signature for the worm. This happens successfully if the number of observed outbound propagations of the worm exceeds `PRM_SIG_MINSTREAMS`:

```
void outboundTrafficNotification() {
  // Does the traffic "slip past us"?
  if (Chance.occurs(PRM_PROB_TRAFFIC_OBSERVED)) {
```

```
// outbound traffic has been observed!
_seentraffic = true;

int numStreams = 0;
int totalPeers = 0;
// begin Rabinizing...
for (Iterator it=_peerHoneypots.iterator(); it.hasNext
  ();)
{
  Honeypot h = (Honeypot)it.next();
  if (h._seentraffic) numStreams++;
  totalPeers++;
}

if (numStreams >= PRM_SIG_MINSTREAMS) {
  // signature construction succeeds
  _antidote = true;
  if (diagOn()) {
    System.out.println("Antidote developed on "+this.
      name());
  }
  if (PRM_SELFISH_HONEYPOT) {
    // the admin takes the honeypot offline
    this.suicide();
  }
}
}
}
}
```

The simulation is made more sophisticated by introducing the possibility that the worm may manage to transmit itself undetected (with probability PRM_PROB_ TRAFFIC_OBSERVED). It is also possible that the administrator of the honeypot, seeing that it is infected and has developed a signature, will take the honeypot offline and refuse to share the antidote with others in the honeypot community.

3.3.3 System Initialization

Env is a singleton class:

```
private static Env_instance;
public static Env instance() {
  if (_instance == null) {
```

```
  _instance = new Env();
}
return _instance;
}
```

In its constructor, `Env` sends itself a `Boot` event:

```
private Env() {
  super();
  Boot b = new Boot();
  this.send(this, b, 0.0);
}
```

`Env` is responsible for instantiating the network of machines, sensors, and honeypots, linking them up at the beginning of the simulation, and introducing the first infection into the network:

```
public void recv(SimEnt src, Event ev) {
  if (ev instanceof Boot) {
    this.initialize();
    // the initial infection!
    Machine m = this.getRandomMachine();
    System.out.println("# Initial infection on: "+m.name
      ());
    m.infect();
  }
}
```

```
public void initialize() {
  // make machines
  for (int i=0; i<PRM_NUM_MACHINES; i++) {
    this.addMachine();
  }
  // make honeypots
  for (int i=0; i<PRM_NUM_HONEYPOTS; i++) {
    this.addHoneypot();
  }
  // make sensors
  for (int i=0; i<PRM_NUM_SENSORS; i++) {
    this.addSensor();
  }
```

```
  // link it up
  this.linkMachines();
  this.linkHoneypots();
  this.linkSensors();

  if (diagOn()) {
    System.out.println("** Initial Environment **\n"+this.
      toString());
  }
}
```

Env maintains data structures for the machines, sensors, and honeypots:

```
private final LinkedList _machines = new LinkedList();
private final LinkedList _honeypots = new LinkedList();
private final LinkedList _sensors = new LinkedList();
```

It supports methods to add to and remove from these data structures. These are used in the implementation of the `initialize()` method:

```
public void addMachine() {
  Machine m = new Machine();
  _machines.add(m);
}

public void addHoneypot() {
  Honeypot h = new Honeypot();
  _honeypots.add(h);
}

public void addSensor() {
  Sensor s = new Sensor();
  _sensors.add(s);
}
```

Env also provides methods to get random elements from these three categories:

```
public Machine getRandomMachine() {
  if (_machines.size() == 0) return null;
  int index = (int)(Math.random()*(double)_machines.size());
  return (Machine)_machines.get(index);
```

```
}

public Honeypot getRandomHoneypot() {
  if (_honeypots.size() == 0) return null;
  int index = (int)(Math.random()*(double)_honeypots.size());
  return (Honeypot)_honeypots.get(index);
}

public Sensor getRandomSensor() {
  if (_sensors.size() == 0) return null;
  int index = (int)(Math.random()*(double)_sensors.size());
  return (Sensor)_sensors.get(index);
}
```

The network topology is formed by having Env tell each machine to link itself up:

```
public void linkMachines() {
  for (Iterator it=_machines.iterator(); it.hasNext();) {
    Machine m = (Machine)it.next();
    m.linkUp();
  }
}

public void linkHoneypots() {
  for (Iterator it=_honeypots.iterator(); it.hasNext();) {
    Honeypot h = (Honeypot)it.next();
    h.linkUp();
  }
}

public void linkSensors() {
  for (Iterator it=_sensors.iterator(); it.hasNext();) {
    Sensor s = (Sensor)it.next();
    s.linkUp();
  }
}
```

A machine links itself up by querying Env for other random machines, which it then designates as its peers:

```
void linkUp() {
  // link up to other machines
  for (int i = 0; i < PRM_NUM_PEER_MACHINES; i++) {
    Machine peer = Env.instance().getRandomMachine();
    if ((peer!=null) && (peer != this)) {
      addPeer(peer);
    }
  }
}

void addPeer(Machine h) {
  if (!_peerMachines.contains(h)) {
    _peerMachines.add(h);
    h.addPeer(this);
  }
}
```

Note in the above code that for addPeer(), the peer "h" is requested to add this machine as its peer. Even though communication is not bidirectional (the worm never needs to move backward whence it came), the "peer" relationship status must be kept by both machines, since machines are allowed to be taken offline during the simulation. In the event of a machine being taken offline, its peers need to be notified so that they can replace it by choosing another peer (at random).

In contrast, during linkUp(), a sensor asks Env for a random set of honeypots with which to connect itself:

```
void linkUp() {
  // Link up w/ Honeypots
  for (int i = 0; i < PRM_NUM_FEED_HONEYPOTS; i++) {
    Honeypot peer = Env.instance().getRandomHoneypot();
    if (peer!=null) {
      addPeer(peer);
    }
  }
  // Link up with the network
  super.linkUp();
}

void addPeer(Machine h) {
  // Keep track of Honeypot peers separately
```

```
  if (h instanceof Honeypot) {
    if (!_feedHoneypots.contains(h)) {
      _feedHoneypots.add(h);
      h.addPeer(this);
    }
  }
  else super.addPeer(h);
  }
```

This is the similar to what a honeypot does when it is asked to undertake `linkUp()`:

```
void linkUp() {
  // A Honeypot's peers are only other Honeypots and
      Sensors...
  // Connect to Honeypots...
  for (int i = 0; i < PRM_NUM_PEER_HONEYPOTS; i++) {
    Honeypot peer = Env.instance().getRandomHoneypot();
    if ((peer!=null) && (peer != this)) {

      addPeer(peer);
    }
  }
  // The Sensors are responsible for linking to Honeypots...
  // not the other way round, so we are done here...
}
```

```
void addPeer(Machine h) {
  // A Honeypot's peers are only other Honeypots or Sensors
  if (h instanceof Honeypot) {
    if (!_peerHoneypots.contains(h)) {
      _peerHoneypots.add(h);
      h.addPeer(this);
    }
  }
  else if (h instanceof Sensor) {
    if (!_feedSensors.contains(h)) {
      _feedSensors.add(h);
      h.addPeer(this);
    }
  }
}
```

Env, having access to all the machines, sensors, and honeypots, is in a position to be able to determine what percentage is infected, both in the network as a whole and in the community of honeypots:

```
public double percentageInfected_Honeypots() {
  int num=0;
  int den=0;
  for (Iterator it=_honeypots.iterator(); it.hasNext();) {
    Honeypot h = (Honeypot)it.next();
    if (h.isInfected()) num++;
    den++;
  }
  return ((double)num/(double)den);
}

  public double percentageInfected_Machines() {
  int num=0;
  int den=0;
  for (Iterator it=_machines.iterator(); it.hasNext();) {
    Machine m = (Machine)it.next();
    if (m.isInfected()) num++;
    den++;
  }
  return ((double)num/(double)den);
}
```

Within the community of honeypots, it can determine the percentage that have developed antidotes or signatures:

```
public double percentageDiagnosed_Honeypots() {
  int num=0;
  int den=0;
  for (Iterator it=_honeypots.iterator(); it.hasNext();) {
    Honeypot h = (Honeypot)it.next();
    if (h.isDiagnosed()) num++;
    den++;
  }
  return ((double)num/(double)den);
}
```

3.3.4 Performance Measures

The `Experimenter` periodically (via `TimeTick` events) checks to see what percentage of the network is infected, what percentage of the honeypots have been infected, and what percentage of the honeypots have developed an antidote. It does this by using the appropriate three methods of the `Env` class (described above). The `Experimenter` notes whenever these three percentages cross 50%. The simulation is considered sufficiently "advanced" (temporally) when more than half the machines and more than half the honeypots are infected. The assumption is that by the time half the honeypots are infected, almost any honeypot operator can develop an antidote. At this point, the `Experimenter` outputs two critical statistics:

- **Infection:** What percentage of machines in the network were infected at the time when half the honeypots were infected?
- **Lead:** What is the ratio of time between when half the honeypots were infected and half the machines were infected?

For example, when we run the program, we might see:

```
infection=4.0  lead=34.78260869565217
```

which would mean that only 4% of the network was infected at the moment when half the honeypots were infected, and that this moment occurred in 30% of the total time it required to infect the majority of the machines.

When more than half the honeypots have developed antidotes, the simulation ends—this is achieved by having the `Experimenter` send itself a `Kill` event, which upon receipt stops `Scheduler`. Below is the code for the entire `recv()` method of the `Experimenter`:

```java
public void recv(SimEnt src, Event ev) {
  // dispatch!
  if (ev instanceof Kill) {
    this.suicide();
    Scheduler.instance().stop();
  }
  else if (ev instanceof TimeTick) {
    TimeTick tt = (TimeTick)ev;
    if (_virgin) {
      System.out.println("# time \t %mach-inf \t" +
        " %hp-inf \t % hp-diag");
      _virgin = false;
      _lastpim = _pim = _env.percentageInfected_Machines();
```

```
  _lastpih = _pih = _env.percentageInfected_Machines();
  _lastpdh = _pdh = _env.percentageDiagnosed_Honeypots
    ();
}
else {
  _lastpim = _pim;
  _lastpih = _pih;
  _lastpdh = _pdh;

  _pim = _env.percentageInfected_Machines();
  _pih = _env.percentageInfected_Honeypots();
  _pdh = _env.percentageDiagnosed_Honeypots();
}

double time = Scheduler.instance().getTime();
// determine stats

if ((_lastpih < 0.5) && (_pih >= 0.5)) {
  _pihThresholdTime = time;
  _pihThreshold = _pim;
}
if ((_lastpim < 0.5) && (_pim >= 0.5)) {
  _pimThresholdTime = time;
  _pimThreshold = _pim;
}
if ((_lastpdh < 0.5) && (_pdh >= 0.5)) {
  _pdhThresholdTime = time;
  _pdhThreshold = _pim;
}

if (diagOn()) {
  System.out.println(""+time+
    "\t"+_pim+
    "\t"+_pih+
    "\t"+_pdh);
}

if ((_pim >= 0.5) && (_pih >= 0.5)) {
  this.send(this, new Kill(), 0);
  System.out.println("infection="+(100.0*_pihThres
    hold)+"\tlead="
```

```
            (100.0*_pihThresholdTime/_pimThresholdTime));
    }
    // register next tick
    this.send(this, tt, tt.getPeriod());
  }

}
```

3.3.5 System Parameters

In the course of defining the various `SimEnts`, we have made use of many parameters. Let us catalog them now:

- `Machine` (an infectable computer or network element):
  - PRM_NUM_PEER_MACHINES
  - PRM_PROB_VULNERABLE
- `Honeypot` (a special kind of `Machine`):
  - PRM_NUM_PEER_HONEYPOTS
  - PRM_PROB_TRAFFIC_OBSERVED
  - PRM_SIG_MINSTREAMS
  - PRM_SELFISH_HONEYPOT
- `Sensor` (another special kind of `Machine`):
  - PRM_SENSOR_WAKEUP
  - PRM_PROB_INFECTION
  - PRM_NUM_FEED_HONEYPOTS
- `Env` (a network of `Machines`):
  - PRM_NUM_MACHINES
  - PRM_NUM_HONEYPOTS
  - PRM_NUM_SENSORS
- `Worm` (the malicious software):
  - PRM_MAX_INSTANCES
  - PRM_WORM_WAKEUP

Rather than have these parameters take values that are set in each of the individual class source files, we can aggregate them all into a single class called `Param`, so that there is a centralized place where simulation parameters are defined:

```
// Environment
// number of honeypots
public static int    Env_PRM_NUM_HONEYPOTS = 8;
```

```
// number of sensors
  public static int     Env_PRM_NUM_SENSORS = 10;

  // number of machines that are neither honeypots, nor sensors
  public static int     Env_PRM_NUM_MACHINES =
Env_PRM_NUM_SENSORS*100;

  // Machines

  // each machine connects to this many other machines
  public static int     Machine_PRM_NUM_PEER_MACHINES = 4;

  // probability that a machine is vulnerable
  public static double Machine_PRM_PROB_VULNERABLE = 0.99;

  // Sensors

  // a sensor sees spontaneous traffic every this many seconds
  public static double Sensor_PRM_SENSOR_WAKEUP = 5.0;

  // the probability that a sensor sees a worm in spontaneous
     traffic
  public static double Sensor_PRM_PROB_INFECTION = 0.0;

  // each sensor feeds this many honeypots
  public static int Sensor_PRM_NUM_FEED_HONEYPOTS = 2;

  // Honeypots

  // each honeypot attempts to connect to this many other
     honeypots
  public static int

Honeypot_PRM_NUM_PEER_HONEYPOTS=Env_PRM_NUM_HONEYPOTS/2;

  // the probability that outbound traffic is detected as being
     a worm
  public static double Honeypot_PRM_PROB_TRAFFIC_OBSERVED =
     1.0;
  // how many streams are needed to construct the filter
     signature
```

```
public static int      Honeypot_PRM_SIG_MINSTREAMS=
  Honeypot_PRM_NUM_PEER_HONEYPOTS/2;
// is the honeypot selfish?
// That is, it disconnects after discovering the antidote?
public static boolean Honeypot_PRM_SELFISH_HONEYPOT =
  false;

// Worm

// can multiple instances of the worm exist on a given
    honeypot?
// If so how many at most?
public static int      Worm_PRM_MAX_INSTANCES = 1;

// the worm makes up every this many seconds
public static double Worm_PRM_WORM_WAKEUP = 5.0;
```

The `Param` class can also then include methods to read in the values of these parameters from a file (or save them to a file). The description of how these are done is not explained here since it is not really essential to the understanding of developing discrete-event simulations.

3.3.6 The Events

The majority of the logic for the simulation lies in the `SimEnts`. `Event` subclasses do not need to mutate on delivery into a `SimEnt`. This makes their code very simple. There are four types of events: the `Boot` event is used by `Env` to trigger the creation of the network, and the `Kill` event is used by the `Experimenter` to trigger the end of the simulation. The `TimeTick` event is used to maintain the life cycle of the worm instances, and to allow the `Experimenter` to make periodic measurements of the system status (to determine if the simulation can be stopped). The `Infection` event carries the worm's malicious code from an infected machine to one of its random peers. Below is the code for the four concrete `Event` subclasses:

```
public class Boot implements Event {
  public Boot() {
  }

  public void entering(SimEnt locale) {
    // no op
  }
}
```

```
public class Kill implements Event {
  public Kill() {
  }
  public void diag() {
    System.out.println("Kill: "+Scheduler.instance().
      getTime());
  }

  public void entering(SimEnt locale) {
    // no-op
  }
}

public class TimeTick implements Event {
  private final double _period;

  public TimeTick(double period) {
    _period = period;
  }

  public void entering(SimEnt locale) {
    // no-op
  }

  public double getPeriod() {
    return _period;
  }
}

public class Infection implements Event {
  public Infection() {}
  public String toString() {
    return ("Infection@"+Scheduler.instance().getTime());
  }

  public void entering(SimEnt locale) {
    // no-op
  }
}
```

3.4 Simulation Execution

The simulation proceeds in three steps: initialize the parameters, make `Env` (network), and make the `Experimenter`. Then let `Scheduler` run. Underneath, `Env`, upon being instantiated, initiates sending a `Boot` event to itself, which causes the machines, sensors, and honeypots to be instantiated and linked up. `Env` also infects a randomly chosen machine. The simulation is underway. Below is the code for the `main()` method of the simulation:

```java
public class Main {
  public static void main(String[] args) {
    try {
      System.out.println("# Starting Scheduler...");
      Param.initialize();
      Env env = Env.instance();
      Experimenter exp = new Experimenter(env, 1.0);
      Thread t = new Thread(Scheduler.instance());
      t.start();
      try { t.join(); }
      catch (Exception e) { }

      System.out.println("# ...Scheduler exiting.");
    }
    catch (Exception ex) {
      ex.printStackTrace();
    }
  }
}
```

Now we can compile the code:

```
javac -g -d ./classes -classpath ../../classes/fw.jar *.java
  */*.java
```

When we run it, we get:

```
# Starting Scheduler...
# Env_DIAG=false
# Machine_DIAG=false
# Honeypot_DIAG=false
# Sensor_DIAG=false
```

```
# Worm_DIAG=false
# Scheduler_DIAG=false
# Experimenter_DIAG=false
# Env_PRM_NUM_HONEYPOTS=4
# Env_PRM_NUM_SENSORS=5
# Env_PRM_NUM_MACHINES=50
# Machine_PRM_NUM_PEER_MACHINES=3
# Machine_PRM_PROB_VULNERABLE=0.99
# Sensor_PRM_SENSOR_WAKEUP=5.0
# Sensor_PRM_PROB_INFECTION=0.0
# Sensor_PRM_NUM_FEED_HONEYPOTS=1
# Honeypot_PRM_NUM_PEER_HONEYPOTS=2
# Honeypot_PRM_PROB_TRAFFIC_OBSERVED=1.0
# Honeypot_PRM_SIG_MINSTREAMS=2
# Honeypot_PRM_SELFISH_HONEYPOT=false
# Worm_PRM_MAX_INSTANCES=1
# Worm_PRM_WORM_WAKEUP=5.0
# Initial infection on: M59
# time   %mach-inf  %hp-inf  % hp-diag
infection=4.0  lead=34.78260869565217
# ...Scheduler exiting.
```

3.5 Output Analysis

The previous execution section shows two performance measures: (1) the infection in the ambient network was only 4% at the time when more than half the honeypot machines succeeded in constructing an antidote; (2) the time at which this event took place was 34% of the total time that the malware required to capture the majority of the ambient network. Together, these performance measures indicate that the honeypot community would have succeeded in developing an antidote well in advance of the period during which the infection spread most virulently (i.e., when the percentage of machines infected exploded from 4% to over 50%).

In the sensitivity analysis stage of the simulation study, we ran the experiments many times, varying the system parameters to determine how they impact the infection and lead time performance measures. Recall from the list in Section 3.5.5 that the simulation study had many parameters. These parameters were taken to have plausible values, but the extent to which their values influence the two performance measures (and the nature of this influence) must be determined by systematic repeated simulations in which the parameter values are perturbed in order to determine and quantify the effect.

3.6 Summary

We developed a discrete-event simulation of a honeypot community-based system for automatic worm detection and immunization. The simulations showed that the infection in the ambient network was lower than 5% at the time when most honeypot machines were able to construct an antidote. This antidote construction became possible early, in about one-third of the time required for the malware to infect the majority of ambient networks. The simulations indicate that the proposed system architecture is effective at antidote generation for emergent malware. This is just one application of many where discrete-event simulation could be used. The reader could extend this type to other network applications, such as P2P, sensor networks, ad hoc networks, collaborative networks, cognitive radios, multimedia over wireless, as well as Internet applications.

Recommended Reading

[1] J. Sandin and B. Khan, "Petrifying worm cultures: scalable detection and immunization in untrusted environments," *Proceedings of IEEE International Conference on Communications (ICC 2007), Glasgow, Scotland*, IEEE, 2077, pp. 1423–1428.

[2] V. Paxson, "Bro: a system for detecting network intruders in real-time," *Computer Networks*, vol. 31, no. 23–24, pp. 2435–2463, 1999.

[3] D. Moore, V. Paxson, S. Savage, C. Shannon, S. Staniford, and N. Weaver, "Inside the slammer worm," *IEEE Security and Privacy*, vol. 1, no. 4, pp. 33–39, 2003.

[4] C. C. Zou, L. Gao, W. Gong, and D. Towsley, "Monitoring and early warning for internet worms," *Proceedings of the 10th ACIv1 Conference on Computer and communications security*, ACM Press, 2003, pp. 190–199.

[5] C. C. Zou, W. Gong, and D. Towsley, "Worm propagation modeling and analysis under dynamic quarantine defense," *Proceedings of ACM Workshop on Rapid Malcode*, ACM Press, 2003, pp. 51–60.

[6] N. Weaver, S. Staniford, and V. Paxson, "Very fast containment of scanning worms," *Proceedings of the 13th USENIX Security Symposium*, USENIX Association, 2004.

[7] H.-A. Kim and B. Karp, "Autograph: toward automated, distributed worm signature detection," *Proceedings of the 13th USENIX Security Symposium*, USENIX Association, 2004.

[8] C. Kreibich and J. Croweroft, "Honeycomb: creating intrusion detection signatures using honeypots," *SIGCOMM Computer Communication Review*, vol. 34, no. 1, pp. 51–56, 2004.

[9] B. Dragovic, K. Fraser, S. Hand, T. Harris, A. Ho, I. Pratt, A. Warfield, P. Barham, and R. Neugebauer, "Xen and the art of virtualization," *Proceedings of the ACM Symposium on Operating Systems Principles*, ACM, 2003.

[10] I. Stoica, R. Morris, D. Karger, F. Kaashoek, and H. Balakrishnan, "Chord: a scalable peer-to-peer lookup service for internet applications," *Proceedings of ACM SIGCOMM*, ACM, 2001, pp. 149–160.

4

Monte Carlo Simulation

Monte Carlo is a classical simulation technique. In fact, Monte Carlo is only one particular application of an otherwise general method, which is applicable in both deterministic and probabilistic settings. At the heart of Monte Carlo is a computational procedure in which a performance measure is estimated using samples drawn randomly from a population with appropriate statistical properties. The selection of samples, in turn, requires an appropriate random number generator (RNG). Ideally, the generated "random" sequences are a completely faithful software counterpart of the non-determinism underlying the actual process.

The term *Monte Carlo method* is generally used to refer to any simulation techniques related to the use of random numbers. This includes methods such as Monte Carlo simulation, Monte Carlo integration, importance sampling, genetic algorithms, simulated annealing, the Hasting–Metropolis algorithm, percolation, random walk, and ballistic deposition, just to name a few. Of these, *Monte Carlo simulation* is perhaps the most common and well-understood class of numerical methods for computer simulations and experiments. Monte Carlo simulation generates random inputs (in reference to a known distribution) and uses the inputs to determine a numerical value of performance measure.

The essential part of the Monte Carlo simulation is a procedure for generating a pseudo random number R that is distributed uniformly over the interval $0 < R < 1$. Such a procedure is applied repeatedly within the Monte Carlo simulation to produce a sequence of independently generated random numbers.

4.1 Characteristics of Monte Carlo Simulations

The input of the simulation is a distribution (based on some theoretical model of reality, or one built empirically from observations) over the set of inputs. Based on this

Network Modeling and Simulation   M. Guizani, A. Rayes, B. Khan and A. Al-Fuqaha
© 2010 John Wiley & Sons, Ltd.

distribution, the Monte Carlo simulation samples an instance of the inputs and runs the simulation for this specific instance. The previous step of selecting inputs and running the simulation is performed many times, and the results of these multiple simulations are then aggregated in a numerically meaningful manner. The three features common to all Monte Carlo simulations are:

1. A known distribution over the set of system inputs.
2. Random sampling of inputs based on the distribution specified in feature 1, and simulation of the system under the selected inputs.
3. Numerical aggregation of experimental data collected from multiple simulations conducted according to feature 2.

Numerical experiments of Monte Carlo simulation lead us to run the simulation on *many sampled inputs* before we can infer the values of the system performance measures of interest.

4.2 The Monte Carlo Algorithm

The Monte Carlo algorithm is based on the so-called *law of large numbers*. This mathematical result of probability theory asserts that if one generates a *large number N* of samples $x_1, x_2, \ldots, x_N$ from a space X, and compute a function f of each of the samples, $f(x_1), f(x_2), \ldots, f(x_N)$, then the mean of these values will *approximate* the mean value of f on the set X. That is, as N tends to $|X|$, we see that the following holds:

$$(1/N)[f(x_1)+f(x_2)+ \cdots +f(x_N)] \sim f*(X) \text{ the mean value of } f \text{ on the elements of } X.$$

4.2.1 A Toy Example: Estimating Areas

To further understand Monte Carlo simulation, let us examine a simple problem. Below is a rectangle for which we know the length (10 units) and height (4 units). It is split into two sections which are identified using different colors. What is the area covered by the dark area?

Due to the irregular way in which the rectangle is split, this problem is not easily solved using analytical methods. However, we can use Monte Carlo simulation to easily find an approximate answer. The procedure is as follows:

1. Randomly select a location within the rectangle.
2. If it is within the dark area, record this instance as a hit.
3. Generate a new location and repeat 10 000 times.

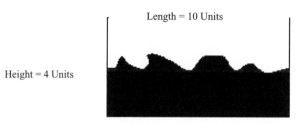

What is the area under the curve (covered by the dark area)?

Figure 4.1   Distribution comparison

After using Monte Carlo simulation to test 10 000 random scenarios, we will have a pretty good average of how often the randomly selected location falls within the dark area.

We also know from basic mathematics that the area of the rectangle is 40 square units (length × height). Thus, the dark area can now be calculated by:

$$\text{Dark area} = \frac{\text{number of hits}}{10\,000\ \text{scenarios}} \times 40\ \text{square units.}$$

Naturally, the number of random scenarios used by the Monte Carlo simulation does not have to be 10 000. If more scenarios are used, an even more accurate approximation for the dark area can be obtained.

This almost toy-like example illustrates all the core ideas of Monte Carlo simulation. As we move to a more complicated example, that of a charging car station, the concepts we have seen will carry over:

- The notion of choosing a two-dimensional (2-D) point at random will generalize to a higher-dimensional analog, e.g., choosing a sequence of electric car arrivals.
- The question of whether the randomly chosen point is black or white will generalize from a binary measurement to a continuous measurement, e.g., for each "point" the percentage of time that the attendant of the charging station for electric cars was busy.
- The question of wanting to estimate the area of the black region generalizes to the question of wanting to estimate the average value of the continuous measurement (the average percentage of time that the attendant was busy) across all possible sequences of electric car arrivals. Just as in the toy example above, this cannot be easily determined precisely since we cannot possibly try all possible sequences of electric car arrivals. Fortunately, however, as in the toy example, Monte Carlo simulation lets us estimate this value.

4.2.2 The Example of the Electric Car Battery Charging Station

Let us reconsider the charging station example from Section 1.3.1 of Chapter 1 where we were interested in several performance measures, including:

- f: the maximum size of Q over the simulation's duration; and
- g: the average percentage of time the server is busy.

By design, the simulation had several system parameters, including:

- D (governs electric car inter-creation intervals);
- F (governs the battery capacity); and
- M (the maximum battery charging rate).

We noted that even for fixed values of the system parameters D, F, and M, the performance measures depend heavily on random choices made in the course of the simulation. Thus, the system is non-deterministic and the performance measures are themselves random variables. This is precisely a setting in which Monte Carlo simulation could be applied. We assume that D, F, and M are fixed (by consulting domain experts or by mining real-world trace data from operating charging stations), then we consider X to be the space of all possible electric car arrival histories, for which temporal sequencing is governed by the distribution implicit in D (and battery capacity is governed by the distribution implicit in F). The act of sampling x_1 from X amounts to specifying a single electric car arrival history. For example, the arrival history x_1 might be:

Electric car 1 arrived at time 1.0 having a battery capacity 10 kW that was 10% full.
Electric car 2 arrived at time 2.5 having a battery capacity 14 kW that was 80% full.
Electric car 3 arrived at time 5.5 having a battery capacity 3 kW that was 25% full.
Electric car 4 arrived at time 12.1 having a battery capacity 22 kW that was 5% full.

Now, given this sample, the system's inputs are fully specified. We can run the simulation and determine $f(x_1)$ and $g(x_1)$. The precise values of these performance measures are highly dependent on the system parameters—in this case, the most influential parameter is M (the other parameters D and F influence the electric car arrival history). For some values of M, we might for example find that the maximum queue size $f(x_1) = 2$ and that the server was busy $g(x_1) = 75\%$ of the time. Now, we can select another electric car arrival history x_2 from the space X. For example, the electric car arrival history x_2 might be:

Electric car 1 arrived at time 1.0 having a battery capacity 8 kW that was 20% full.
Electric car 2 arrived at time 2.0 having a battery capacity 18 kW that was 60% full.

Electric car 3 arrived at time 4.5 having a battery capacity 30 kW that was 45% full.
Electric car 4 arrived at time 8.7 having a battery capacity 2 kW that was 95% full.

As before, we can run the simulation and determine $f(x_2)$ and $g(x_2)$. The
procedure is repeated in this manner for a large number of sampled inputs N.
Each of the inputs $x_1, x_2, \ldots, x_N$ is being sampled (i.e., chosen) according to known
and agreed-upon "distributions" that govern how likely each of the inputs (i.e.,
electric car arrival histories) is in practice. In all but the simplest cases, not all the x_i
are deemed to be equally likely—indeed if they are, the system may be amenable to
analytical modeling. Most computer programming languages or spreadsheets have
built-in functions to generate uniform distribution random numbers that can be
easily used. For example, one can use the RAND() function in Microsoft Excel or
RND in Visual Basic to generate random numbers between 0 and 1. One technical
hurdle that must be addressed when implementing a Monte Carlo simulation is how
one can generate other arbitrary (but specified) distributions using only the ability
to generate random numbers uniformly distributed between 0 and 1. Typically, this
feature is provided in most simulation frameworks, via the so-called inverse CDF
method.
   Having collected a large number of performance measures for the sampled inputs
$f(x_1), f(x_2), f(x_3), \ldots, f(x_N)$ and $g(x_1), g(x_2), g(x_3), \ldots, g(x_N)$, we invoke the law of
large numbers to assert that the quantity:

$$(1/N)[f(x_1) + f(x_2) + \cdots + f(x_N)]$$

approaches $f^*(X)$, the mean value of f on the elements of X; and the quantity:

$$(1/N)[g(x_1) + g(x_2) + \cdots + g(x_N)]$$

approaches $g^*(X)$, the mean value of g on the elements of X. How large N needs to be for
the "approach" to be close enough is often a difficult question to answer. We will see
how to incorporate determining N's value in the course of the simulation itself.
Informally, the simulation will repeatedly sample the next x_i from X, compute $f(x_i)$ and
$g(x_i)$, and then ask "have we generated enough samples to be sure our estimate of the
performance is good?" Exactly this will be implemented computationally and will be
illustrated when we implement the electric car charging station example, using our
discrete-event simulation framework developed and discussed in Chapter 2.

4.2.3 Optimizing the Electric Car Battery Charging Station

Now, let us formulate an optimization problem for this system of the electric car
charging station. Suppose we find through surveys that customers would be willing to

join a line of four electric cars—but if the line at the station is five or more electric cars in length, then customers would rather look for a different charging station. Management is then very motivated to make sure that there are never more than four electric cars in line. Suppose the electric charger manufacturer can provide chargers with arbitrary high charging rates, but chargers with high charging rate cost more because insurance costs are higher due to the likelihood of accidents. Suppose that a charger with a charging rate of M kW per minute would cost the charging station's owner $\$1000M^2$. We know that D governs the electric car inter-creation intervals and F governs battery capacity. To minimize costs and maximize profits, we seek the smallest value of M (charging rate) for which the maximum queue size exhibited by the system is four or lower. How could this magical value of M be determined?

The essence of the approach is to set M to some number m_1, run the Monte Carlo simulation to determine the performance measure $f^*(X)$ given that we have set $M = m_1$. Since we intend to vary the value of M we will write $f^*(X; m_1)$ in place of $f^*(X)$ so that the notation reminds us that this is the value of f when $M = m_1$. Now, intuitively, we know that as the value of M is increased, $f^*(X)$ will decrease, since electric cars can be serviced very quickly. We will increase M incrementally until we see that $f^*(X)$ has dropped to a value that is at or below 4. If we seek the minimal value of M for which this is true, then we will have to perform a binary search on M to find the precise value at which $f^*(X)$ crosses 4.

4.3 Merits and Drawbacks

The main advantages of using a Monte Carlo simulation are:

- It provides approximate solutions to many mathematical problems concerning systems that involve an element of randomness that can be modeled via a distribution.
- It also provides a framework for statistical sampling of inputs, allowing results from numerical experiments to be aggregated in a meaningful way.
- Its use is quite simple and straightforward as long as convergence can be guaranteed by the theory.
- For optimization problems, it often can reach global optima and overcome local extrema.

A Monte Carlo simulation has the following limitations:

- It was originally developed to study properties of stable systems that are in equilibrium—systems for which the values of performance measures are not fluctuating. This means that X must be constructed in such a way that each sample

x_i determines a system in steady state. In our previous exposition, we took X to be the set of arrival sequences for four electric cars. However, a charging station is probably not in a "steady state" upon handling a sequence of four electric cars; its performance measures f and g will not have stabilized. It is more likely to be in a steady state (i.e., its performance measures f and g would have stabilized) when it has seen a large number of electric cars, perhaps 40 000. We will see how to deal with the system if it is in steady state and how to incorporate that into the simulation itself. Informally, the simulation will construct samples from X and assess whether the sample describes a system in steady state or not. If it is a sample of length 4, the answer will probably be no, in which case it will be extended as needed. This will continue until eventually, when it is a sample of 40 000 electric car arrivals, the simulation will determine that the system defined by the sample is indeed in steady state. How exactly this is implemented computationally will be illustrated when we implement the charging station as a simulation, using the discrete-event simulation framework developed in Chapter 2.

- It is not universally agreed whether this method can also be used to simulate a system that is not in equilibrium (i.e., in a transient state).
- To use a Monte Carlo simulation, one needs to generate a large number of input samples—since before doing so, the law of large numbers need not to hold. This can be time consuming in practice if each simulation itself takes a long time.
- The results of a Monte Carlo simulation are only an *approximation* of the true values.
- When the performance measure has a high variance, the values produced by a Monte Carlo simulation will converge to the mean value over X, but any given sample from X may vary greatly from the computed mean.
- The distribution over the set of inputs must be known in order to sample from X.

4.4 Monte Carlo Simulation for the Electric Car Charging Station

In this section we will design and implement a Monte Carlo simulation for our example of an electric car charging station, using the discrete-event simulation framework developed in Chapter 2.

The entities in the simulation are:

- `TrafficGen`
- `ChargingStation`
- `Server`

Let us design and implement each of these in turn.

4.4.1 The Traffic Generator

The first step is to put together the `TrafficGen` class. The main purpose of `TrafficGen` is to act as a factory for `Car` objects, which, once made, will be delivered to the `ChargingStation` entity by sending it a `CarArrival` event. Here is an empty stub for `TrafficGen`:

```
public class TrafficGen extends SimEnt {
};
```

As data members, one would need `TrafficGen` to know `ChargingStation`:

```
private ChargingStation _gs;
```

and to keep track of how many cars it has made:

```
private int _numCars = 0;
```

The behavior of TrafficGen is determined by two doubles:

```
private double _D;
private double _F;
```

D is the inter-car creation time and F is the maximum battery size. Given these data members, the constructor of `TrafficGen` is easy to write

```
TrafficGen (double intercartime, double maxbatterysize,
  ChargingStation gs) {
  super();
  _D = intercartime;
  _F = maxbatterysize;
  _gs = gs;
  send(this, new MakeCar(), 1.0);
}
```

Note that a `MakeCar` event is made and sent by `TrafficGen` to itself—this event's arrival is what triggers the making of a `Car` (and sending the `Car` to the `Charging-Station` via a `CarArrivalEvent`). We see this in our implementation of `TrafficGen`'s recv method:

```
public void recv (SimEnt src, Event ev) {
```

```
if (ev instanceof MakeCar) {
  Car c = new Car(_F);
  _numCars++;
  CarArrival ca = new CarArrival(c);
  send(_gs, ca, 0.0);
  send(this, ev, waitingTime());
}
}
```

We observe that the receipt of a MakeCar event results in making a Car, incrementing the _numcars variable, and sending the Car (encapsulated in a CarArrival event) to the ChargingStation. The MakeCar event (ev) is then scheduled by TrafficGen to be sent to itself, at some time in the future:

```
private double waitingTime() {
  return _D * Math.random();
}
```

It may be convenient to have a method to get the number of cars made so far:

```
int numCars() {
  return _numCars;
}
```

Every SimEnt requires these two abstract methods to be defined:

```
public String getName() {
  return "TrafficGenerator";
}

public void deliveryAck(Scheduler.EventHandle h) {
  // no op
}
```

We do not need to do anything when we get an acknowledgement that a "sent event" has been delivered, so a null body implementation of deliveryAck() is sufficient.

Let us add a printout line for debugging:

```
if (Run.DEBUG)
  System.out.println("TrafficGenerator: made a car...");
```

For performance measurements to detect when steady state is achieved, we modify `TrafficGen` by adding to the constructor:

```
Run.resetPercPerformanceMeasureStats();
```

then, inside the `recv()` method, when `TrafficGen` makes a car, it does the following:

```
Run.calculatePercPerformanceMeasure();
if (_numCars % 100 == 0) {
  if (! Run.terminatingBecausePercMeasureConverged()) {
    Run.resetPercPerformanceMeasureStats();
  }
  else {
    Scheduler.instance().stop();
  }
}
```

The code fragment above calculates the performance measures and once the variables have stabilized, the system is deemed to have reached a steady state. The operation of the `TrafficGen` class requires two types of events: `MakeCar` and `CarArrival`. These are relatively straightforward in their implementation:

```
public class MakeCar implements Event {
  public void entering(SimEnt locale) {
    // no op
  }
}

public class CarArrival implements Event {
  Car _car;

  CarArrival(Car c) {
    _car = c;
  }

  public void entering(SimEnt locale) {
    // no op
  }
}
```

As is evident from this code, `MakeCar` is just a tagging event class, while `CarArrival` is a container for instances of `Car`. What exactly is `Car`? Let us consider this question next.

4.4.2 The Car

For the purposes of this simulation, `Car` is a partly full battery of charging power. As such, it requires a `capacity` and a `contents` data member:

```
public class Car {
  private double _K; // battery capacity in kWs
  private double _G; // battery contents in kWs
}
```

Since cars are to have batteries of random sizes, we decide that car batteries are uniformly distributed in an interval [0. . .*F*]. This might not be a reasonable assumption; we would need validation of the model from domain experts to assess how reasonable it is. Presuming this, let us proceed with the actual implementation:

```
Car(double F) {
  super();
  _K = F * Math.random();
  double L = 100.0 * Math.random();
  _G = L * _K / 100.0;
}
```

Now, one can see that when `Car` is made, we choose a random value in [0. . .*F*] as its capacity. Then, we choose a random level in [0. . .100]%. From these, we are able to compute the actual amount of power in the battery and store it in the data member _G. When a car arrives and is filled up at a charging station, what happens? The level in the battery _G gets set to the capacity of the battery _K (this is what "battery is charged" means). How long does this operation take? The answer depends on the flow rate being used to fill the battery of the car. If the flow rate is *M* kW per minute, the time required to charge the battery will be (_K-_G)/*M* minutes. Let us define a `fill()` method for `Car` which will charge the battery at a specified `ChargingStation`:

```
double fill(ChargingStation s) {
  double M = s.chargingRate();
  double time = (_K - _G)/M;
  _G = _K;
  return time;
}
```

Note that `Car` must query `ChargingStation` for the charging rate of its battery, and returns the total time in minutes that is required to fill the car's battery. Now, we are ready to look at what a charging station is and how it operates.

4.4.3 The Charging Station

Consider a charging station which operates at a fixed battery charging rate, a server who charges the batteries, and a queue of cars waiting in line. Thus, an initial implementation of a charging station might be:

```
public class ChargingStation extends SimEnt {
  double _M;
  LinkedList _Q = new LinkedList();
  Server _server;
}
```

The constructor of this class can be implemented as:

```
ChargingStation(double chargingRate) {
    super();
  _M = chargingRate;
  _server = new Server(this);
}
```

`ChargingStation` receives `CarArrival` events, sent to it by `TrafficGen`. What does `ChargingStation` do when a `CarArrival` event is received? Clearly, it adds `Car` (contained in the `CarArrival` event) to the end of its queue `_Q`, but what else must it do? If the car is the first one being added to the queue (i.e., the queue was previously empty), then the server needs to be informed that there should be service to a car. If, however, the car is being added to the end of a non-empty queue of cars, then we expect the server to be busy charging the car battery at the head of this queue—and we do not need to inform the server, because we expect the server to notice that there are still more cars left to be serviced:

```
public void recv(SimEnt src, Event ev) {
  if (ev instanceof CarArrival) {
    CarArrival ca = (CarArrival)ev;
    Car c = ca._car;
    synchronized(_server) {
      boolean wakeupServer = false;
      if (_Q.size()==0 && _server.idle()) wakeupServer = true;
```

```
      _Q.addLast(c);
      if (wakeupServer) send(_server, new ServiceCar(), 0.0);
    }
  }
}
```

It may be convenient to have a method to get the battery charge rate and the next car from the front of the queue:

```
public double getchargingRate() {
  return _M;
}

public Car getNextCar() {
  if (_Q.size()==0) return null;
  else return (Car)_Q.removeFirst();
}
```

Every `SimEnt` requires these two abstract methods to be defined:

```
public String getName() {
  return " ChargingStation ";
}

public void deliveryAck(Scheduler.EventHandle h) {
  // no op
}
```

We do not need to do anything when we get an acknowledgement that a sent event has been delivered, so a null body implementation of `deliveryAck()` is sufficient.

Let us add a printout line for debugging inside of `recvc()`:

```
if (Run.DEBUG)
  System.out.println("ChargingStation: car arrived @" +
    Scheduler.getTime());
```

For performance measurements we modify `ChargingStation` by adding the following line to the `recv()` method after we add `Car` to the end of the queue `_Q`:

```
Run.updateQPerformanceMeasure(_Q.size());
```

The operation of the `ChargingStation` class generates one type of event: `ServiceCar`; this is sent to the server. The `ServiceCar` event is relatively straightforward and operates as a tagging class:

```
public class ServiceCar implements Event {
ServiceCar() {
}

public void entering(SimEnt locale) {
  // no op
}
```

Now, let us see what the recipient of the `ServiceCar` event does by turning to consider the server.

4.4.4 The Server

A server is either idle or busy, and works in a charging station. Accordingly, we can start with a class with two data members:

```
public class Server extends SimEnt {
  boolean _idle = true;
  ChargingStation _station;
}
```

The constructor for `Server` would then be:

```
Server(ChargingStation gs) {
  _station = gs;
}
```

`Server` receives `ServiceCar` events from `ChargingStation`. What does `Server` do? Here is a first pass at implementing `recv()`:

```
public void recv(SimEnt src, Event ev) {
  if (ev instanceof ServiceCar) {
    synchronized(this) {
      Car c = _station.getNextCar();
      _idle = false;
      double time = c.fill(_station);
```

```
    send(this, new ServiceCar(), time);
  }
}
```

In this code, `Server` charges the battery and schedules the `ServiceCar` event to be sent to itself once the car's battery has been charged. The problem is that this might have been the last car in the queue, in which case, when the `ServiceCar` event arrives the second time, there will be no car (`c=null`). We need to be robust against this, so here is a second attempt at `recv()`:

```
public void recv(SimEnt src, Event ev) {
  if (ev instanceof ServiceCar) {
    synchronized(this) {
      Car c = _station.getNextCar();
      if (c!=null) {
        if (_idle) {
        _idle = false;
        }
        double time = c.fill(_station);
        send(this, new ServiceCar(), time);
      }
      else {
        _idle = true;
      }
    }
  }
}
```

`ChargingStation` uses an `idle()` method of `Server` to determine if `Server` is idle—if so `ChargingStation` sends `Server` a `ServiceCar` event. This requires us to implement a method called `idle()` in the `Server` class:

```
public boolean idle() {
  return _idle;
}
```

Every `SimEnt` requires these two abstract methods to be defined:

```
public String getName() {
  return "Server";
}
```

```
 public void deliveryAck(Scheduler.EventHandle h) {
   // no op
 }
}
```

We do not need to do anything when we get an acknowledgement that a sent event has been delivered, so a null body implementation of deliveryAck() is sufficient.

To support debugging, we should add some output statements to Server. Also, we add support for performance evaluation measurements by modifying Server, adding lines to the recv() method so that Server reports whenever it is transitioning from idle to busy and busy to idle. This makes the final recv() method:

```
public void recv(SimEnt src, Event ev) {
  if (ev instanceof ServiceCar) {
    synchronized(this) {
      Car c = _station.getNextCar();
      if (c!=null) {
        if (_idle) {
          _idle = false;
          if (Run.DEBUG)
            System.out.println("Server: became not idle @" +
              Scheduler.getTime());
          Run.notIdleNotification(Scheduler.getTime());
        }
        if (Run.DEBUG)
          System.out.println("Server: serviced car @" +
            Scheduler.getTime());
        double time = c.fill(_station);
        send(this, new ServiceCar(), time);
      }
      else {
        _idle = true;
        if (Run.DEBUG)
          System.out.println("Server: became idle @" +
            Scheduler.getTime());
        Run.idleNotification(Scheduler.getTime());
      }
    }
  }
}
```

4.4.5 Putting It All Together

The simulation as a whole is controlled by the `Run` class, which implements the `static main()` function. It should contain `ChargingStation` and `Traffic-Gen`, both of which are made here. Then, the discrete-event simulation can start:

```
public class Run {
  public static boolean DEBUG = false;
  public static void main(String [ ] args) {
    // battery charge rate of the charger is 4 kW per minute
    ChargingStation gs = new ChargingStation(4.0);

    // new car arrives once every [0..6] minutes
    // new car has a battery capacity in the range of [0..20] kWs
    // car's battery is [0..100] percent full
    TrafficGen tg = new TrafficGen(6.0, 20.0, gs);

      Thread t = new Thread(Scheduler.instance());
    t.start();
    try { t.join(); }
    catch (Exception e) { }
  }
}
```

We need to manage performance data being reported by `Server` (whenever it transitions from idle to busy and from busy to idle):

```
static double _last_t = 0.0;
static double _totalIdle = 0.0;
static double _totalNotIdle = 0.0;

static void idleNotification(double t) {
  _totalNotIdle += (t-_last_t);
  _last_t = t;
}
static void notIdleNotification(double t) {
  _totalIdle += (t-_last_t);
  _last_t = t;
}
```

This code maintains two variables which accumulate the total time that `Server` has spent in idle and busy states. Let us augment `main()` by adding final diagnostic output at the very end, once the `Scheduler` thread has stopped:

```
double p = 100.0 * _totalNotIdle / (_totalNotIdle + _
  totalIdle);
System.out.println("Server was busy "+p+"% of the time.");
System.out.println("Maximum Q size was "+_maxQ+" cars.");
System.out.println("Simulation required "+tg.numCars()+"
  cars.");
```

We also need to support the operations of resetPercPerformanceMea-
sureStats, calculatePercPerformanceMeasure, and terminating-
BecausePercMeasureConverged, which are used in the TrafficGen object
to determine if enough cars have been fed into the charging station to consider it to have
reached steady state. To this end, we will define that a steady state has been achieved
when the percentage of time that the server was busy does not fluctuate "too much"
within a given interval of time. Here "too much" will be a parameter that can be set,
which will determine the stringency of requirements that must be met for the system to
be declared to have reached a steady state. Below is a sample implementation of the
three methods in question:

```
static double _minPerc, _maxPerc;
static final double PERC_CONVERGENCE_THRESHOLD = 0.1;
static boolean _percConverged = false;

static void resetPercPerformanceMeasureStats() {
  _minPerc = Double.MAX_VALUE;
  _maxPerc = Double.MIN_VALUE;
}

static void calculatePercPerformanceMeasure() {
  double p = 100.0 * _totalNotIdle / (_totalNotIdle + _
    totalIdle);
  if (p < _minPerc) _minPerc = p;
  if (p > _maxPerc) _maxPerc = p;
}

static boolean terminatingBecausePercMeasureConverged() {
  if (_maxPerc - _minPerc < PERC_CONVERGENCE_THRESHOLD) {
    _percConverged = true;
    return true;
  }
  else return false;
}
```

Recall how these methods of Run were used inside the `TrafficGen` class, where in the constructor we called `resetPercPerformanceMeasureStats`, and inside of the `recv()` method we specified that every time a car is made and sent to `ChargingStation` we should:

```
Run.calculatePercPerformanceMeasure();
if (_numCars % 100 == 0) {
  if (! Run.terminatingBecausePercMeasureConverged()) {
    Run.resetPercPerformanceMeasureStats();
  }
  else {
    Scheduler.instance().stop();
  }
}
```

The net effect of the previously listed code fragments is that the system is deemed to have reached a steady state when the server's percentage of time busy fails to fluctuate more than PERC_CONVERGENCE_THRESHOLD percent over an interval of 100 car arrivals. In addition, the Run class has the following data member and method to keep track of the maximum queue size. The method is called by `ChargingStation` whenever a car is added to its queue:

```
static int _maxQ = 0;

static void updateQPerformanceMeasure(int q) {
  if (q > _maxQ) _maxQ = q;
}
```

4.4.6 Exploring the Steady State

Let us see what happens when we run the program with the Run classes:

```
static double PERC_CONVERGENCE_THRESHOLD = 5.0;
static boolean DEBUG = true;
```

The output of the execution is 885 lines long, shown abridged here:

```
javac -g -d ./classes -classpath ../../classes/fw.jar *.java
java -cp ./classes:../../classes/fw.jar fw.power.Run
```

```
TrafficGenerator: made a car...
ChargingStation: car arrived @1.0
Server: became not idle @1.0
Server: serviced car @1.0
Server: became idle @1.076828814748023
TrafficGenerator: made a car...
ChargingStation: car arrived @3.3174533214640336
Server: became not idle @3.3174533214640336
Server: serviced car @3.3174533214640336
Server: became idle @4.3152408310135915
TrafficGenerator: made a car...
ChargingStation: car arrived @7.937233056456469
Server: became not idle @7.937233056456469
Server: serviced car @7.937233056456469
TrafficGenerator: made a car...
ChargingStation: car arrived @7.974995880329053
Server: serviced car @8.282021474763113
Server: became idle @8.330067415977464
[... 850 lines omitted]
TrafficGenerator: made a car...
ChargingStation: car arrived @586.5054952317727
Server: became not idle @586.5054952317727
Server: serviced car @586.5054952317727
Server: became idle @588.8665719560208
TrafficGenerator: made a car...
ChargingStation: car arrived @589.9732534634641
Server: became not idle @589.9732534634641
Server: serviced car @589.9732534634641
Server: became idle @590.4084805442748
TrafficGenerator: made a car...
Server was busy 40.599987136675736% of the time.
Maximum Q size was 4 cars.
Simulation required 200 cars.
```

Having validated the logs to check that the simulation is operating as intended, we run the same code with DEBUG = false, and get a more manageable output:

```
Server was busy 43.5852490446865% of the time.
Maximum Q size was 2 cars.
Simulation required 200 cars.
```

This output is interpreted as follows. As cars were being serviced at the station, we kept track of the percentage of time that the server was busy (over the history to date). By the time 200 cars had been serviced, we noticed that in the last 100 cars the value of the server's business had varied less than 5% (PERC_CONVERGENCE_THRESHOLD=5.0), so we decided that the system was in a steady state and we stopped the simulation. The percentage of time that the server was busy was approximately 43.6%. What happens if we have a more stringent criterion for the steady state? Suppose we set PERC_CONVERGENCE_THRESHOLD to 1.0 and rerun the simulation. In that case, we find:

```
Server was busy 41.30095210225264% of the time.
Maximum Q size was 4 cars.
Simulation required 600 cars.
```

What happens if we have more stringent criteria for the steady state? Suppose now we set PERC_CONVERGENCE_THRESHOLD to 0.1 and rerun the simulation. We then find:

```
Server was busy 42.138985101913796% of the time.
Maximum Q size was 5 cars.
Simulation required 4500 cars.
```

When PERC_CONVERGENCE_THRESHOLD is 0.01, we get:

```
Server was busy 41.464701454076554% of the time.
Maximum Q size was 6 cars.
Simulation required 37900 cars.
```

When PERC_CONVERGENCE_THRESHOLD is 0.001, we get:

```
Server was busy 41.621161392000076% of the time.
Maximum Q size was 8 cars.
Simulation required 304500 cars.
```

When PERC_CONVERGENCE_THRESHOLD is 0.0001, we get:

```
Server was busy 41.67335379126957% of the time.
Maximum Q size was 9 cars.
Simulation required 2691500 cars.
```

When `PERC_CONVERGENCE_THRESHOLD` is 0.000 01, we get:

Server was busy 41.67017089967508% of the time.
Maximum Q size was 11 cars.
Simulation required 20990500 cars.

There are several things to note here. First, we observe that the percentage of time that the server is busy does converge to the neighborhood of the same value as the definition of the steady state is made more stringent. In other words, even though the system must run for a longer time (larger numbers of cars must pass through the system) when the convergence threshold value is set very small, the value of the performance measure (server busy percentage) is robust in that it converges to a constant that is independent of how long the system has been under simulation. In contrast, the performance measure of maximum queue size does not converge to a constant. Indeed, we would expect maximum queue size to be a monotonically increasing function of the amount of time that the system has been simulated—longer simulations make it more likely that there could be "freak incidents" in which car arrivals occur in such a way as to result in long queues. In this case, all we can hope for is to develop an estimate for maximum queue size that is "a function of simulation duration."

4.4.7 Monte Carlo Simulation of the Station

Having understood the notion of steady state, let us return to the initial case in which the Run class data members have values:

```
static double PERC_CONVERGENCE_THRESHOLD = 5.0;
static boolean DEBUG = false;
```

The output of this is:

Server was busy 40.599987136675736% of the time.
Maximum Q size was 4 cars.
Simulation required 200 cars.

Keep in mind here that the 200 cars used to derive the estimate of 40.59% were in fact a randomly generated sequence of cars (in their arrival times, battery sizes, and chagrining levels). What if we had a different set of 200 cars? Would we get 40.59% again? Most likely we would not get the same result. The above experiment of generating 200 random car arrivals and using *this sequence* to estimate that the server is busy 40.59% of the time is the equivalent of picking a random 2-D point in the 4 × 10 rectangle example of Section 4.2.1 above and asking whether the randomly chosen point is black or white. Here, we chose a random sequence of car arrivals that put the system into steady state (5% variation), and asked for the value of the server business percentage on this sequence. In the above toy example (Section 4.2.1), we asked for a random point in the 4 × 10 rectangle and asked for the value of the color function. So, while 40.59% sounds like a perfectly plausible answer, we must have a sense of how dependent this number is on the particular random sequence of car arrivals. To measure this, we must generate many car arrival sequences, all of which put the system into a steady state (e.g., 5% variance) and ask what the value of the server business function is for all of these different car arrival sequences; specifically, what is the mean value and what is the variance?

To code this, we must add an outer wrapper to our simulation loop in Run's main() method, in order to make it run a certain number of trials (specified by the constant TRIALS). For each trial, it computes the percentage of time that the server was busy, and after all trials are executed, it reports the mean value and the standard deviation of the measures made in the trials.[1] Let us consider the revised code for main() in the Run class:

```
public static void main (String [ ] args) {

   int TRIALS = 10;
   int n = TRIALS;
   double perc [] = new double[n];

   do {

      // battery charging rate is 4 kWs per minute
      ChargingStation gs = new ChargingStation(4.0);

      // new car arrives once every [0..6] minutes
      // new car has a battery capacity in the range of [0..20] kWs
      // car's battery is [0..100] percent full
      TrafficGen tg = new TrafficGen(6.0, 20.0, gs);

      Thread t = new Thread(Scheduler.instance());
      t.start();
      try { t.join(); }
      catch (Exception e) { }
```

---

[1] Referring back to the toy example of Section 4.2.1, the mean value here is the analog of the relative area of the black region (i.e., the area of the black region, normalized by the total area of the 4 × 10 rectangle), and the standard deviation is the error estimate of this area.

```
      double p = 100.0 * _totalNotIdle / (_totalNotIdle + _
         totalIdle);
      System.out.println("Server was busy "+p+"% of the
         time.");
      System.out.println("Maximum Q size was "+_maxQ+"
         cars.");
      System.out.println("Simulation required "+tg.numCars()
         +" cars.");

      n-;
      perc[n] += 100.0 * _totalNotIdle / (_totalNotIdle + _
         totalIdle);

      Scheduler.instance().reset();
      _last_t = 0.0;
      _totalIdle = 0.0;
      _totalNotIdle = 0.0;
   }
   while (n > 0);

   double meanperc = 0.0;
   for (n=0;n<TRIALS;n++) meanperc += perc[n];
   meanperc /= (double)TRIALS;

   double var = 0.0;
   for (n=0;n<TRIALS;n++) var += (perc[n] - meanperc)*(perc
      [n] - meanperc);
   var /= (double)TRIALS;
   double dev = Math.sqrt(var);

   System.out.println("Server was busy "+meanperc+"% of the
      time, "+ "std="+dev);
}
```

When we run this code, we get the following output:

```
javac -g -d ./classes -classpath ../../classes/fw.jar *.java
Note: ChargingStation.java uses unchecked or unsafe
   operations.
Note: Recompile with -Xlint:unchecked for details.
java -cp ./classes:../../classes/fw.jar fw.power.Run
```

```
Server was busy 39.4602804160784% of the time.
Maximum Q size was 3 cars.
Simulation required 200 cars.
Server was busy 41.77209593885758% of the time.
Maximum Q size was 3 cars.
Simulation required 200 cars.
Server was busy 42.53583280696096% of the time.
Maximum Q size was 3 cars.
Simulation required 200 cars.
Server was busy 38.23658252611347% of the time.
Maximum Q size was 3 cars.
Simulation required 300 cars.
Server was busy 40.456579350640794% of the time.
Maximum Q size was 4 cars.
Simulation required 200 cars.
Server was busy 43.29099329492339% of the time.
Maximum Q size was 4 cars.
Simulation required 200 cars.
Server was busy 45.04948133086543% of the time.
Maximum Q size was 4 cars.
Simulation required 200 cars.
Server was busy 35.282865804116376% of the time.
Maximum Q size was 4 cars.
Simulation required 200 cars.
Server was busy 42.571069199527386% of the time.
Maximum Q size was 5 cars.
Simulation required 300 cars.
Server was busy 47.66183655340056% of the time.
Maximum Q size was 5 cars.
Simulation required 200 cars.
```
**Server was busy 41.63176172214844% of the time,
std=3.323358268112881**

Note that the percentages vary in each trial. In addition, some trials require longer to reach steady state (300 cars in the ninth trial versus 200 cars in the tenth trial). The trials are each defined as a simulation which goes on long enough to bring the system to steady state (less than 5% variance in the server business measure over the last 100 car arrivals). Depending on the random sampling, sometimes this requires longer simulations (more cars). The last line of the output shows the mean value (across 10 trials) of the server business, and the standard deviation of the values measured in each of these 10 trials. Note that since each trial made different random choices, it produced different performance measures (e.g., the ninth trial reported server busy 42% of the time, while the tenth trial reported the server busy 47% of the time). Clearly, the values have a non-zero spread, and so the standard deviation will be greater than zero. In fact, we see in the last line that the standard deviation of the performance measures across the 10 trials is 3.32.

Several questions could be posed at this point. The first is "what effect will the number of trials have on the value of the mean and the standard deviation?" Let us see what occurs in practice. If we change `TRIALS` from 10 to 100, the last line of output reads:

`Server was busy 41.422609654685814% of the time,`
  `std=3.0972676894689433`

If we change `TRIALS` from 100 to 1000, the last line of output reads:

`Server was busy 41.68637971516185% of the time,`
  `std=3.012845922105874`

If we change `TRIALS` from 1000 to 10 000, the last line of output reads:

`Server was busy 41.61390895944546% of the time,`
  `std=2.930492611147204`

Note that the mean value of the performance measure does not change much, and the standard deviation decreases only slightly. In particular, it seems unlikely that the standard deviation *would go down to zero as the number of trials increases*. In other words, the spread of performance measures when we run 10 trials is commensurate with the spread of the measures when we run 10 000 trials (modeling these as normal distributions), and not all of it is attributable to the number of trials. The reason for this is that much of the spread in these values comes from the underlying 5% variability that is permitted in the definition of steady state. To illustrate this, let us set `TRIALS` to 1000 and vary the `PERC_CONVERGENCE_THRESHOLD`.

With `TRIALS` set to 1000 and `PERC_CONVERGENCE_THRESHOLD` to 5%, we get:

`Server was busy 41.68637971516185% of the time,`
  `std=3.012845922105874`

With `TRIALS` set to 1000 and `PERC_CONVERGENCE_THRESHOLD` to 1%, we get:

`Server was busy 41.57993260857031% of the time,`
  `std=1.7630972379477647`

With TRIALS set to 1000 and PERC_CONVERGENCE_THRESHOLD to 0.1%, we get:

```
Server was busy 41.64882877029293% of the time,
   std=0.6795750456985769
```

With TRIALS set to 1000 and PERC_CONVERGENCE_THRESHOLD to 0.01%, we get:

```
Server was busy 41.6530111413678% of the time,
   std=0.2355306786101936
```

Note that the mean value of the performance measure does not change much, but the standard deviation of the values recorded approaches *zero* as PERC_CONVERGENCE_THRESHOLD is *decreased*. Once again, this is because the spread of the performance measures is largely attributable to the underlying variability that is permitted in the definition of steady state. A Monte Carlo simulation always involves multiple trials of simulating a system, where within each trial the system must be allowed to reach a steady state. There is always a tradeoff between the stringency of the definition of steady state and the number of trials in the outer loop of Monte Carlo trials. Each of these two factors (stringency of the steady-state definition and the number of Monte Carlo trials) contributes to the variability in the performance measurements, and so care must be taken in setting their values in order to get a statistically significant set of performance measures. In this case, for example, with 1000 trials and a steady-state convergence defined as less than 0.01% variance in server business over 100 car arrivals, we get a set of performance measures centered at 41.65% with a standard deviation of 0.23.

4.5 Summary

This chapter introduced the well-known Monte Carlo simulation technique. The technique applies to both deterministic and probabilistic models to study properties of stable systems that are in equilibrium. A random number generator is used by Monte Carlo to simulate a performance measure drawn from a population with appropriate statistical properties. The Monte Carlo algorithm is based on the law of large numbers with the promise that the mean value of a large number of samples from a given space will approximate the actual mean value of such a space.

To understand Monte Carlo simulation, two examples were introduced: a toy example and a charging station for electric cars. Both examples estimated the average value using a large number of samples. The examples illustrated that the Monte Carlo technique will estimate the described value without considering all possible sequences of arrivals.

Recommended Reading

[1] R. Y. Rubinstein and D. P. Kroese, *Simulation and the Monte Carlo Method*, 2nd Edition, John Wiley & Sons, Inc., 2008.

[2] C. P. Robert and G. Casella, *Monte Carlo Statistical Methods*, 2nd Edition, Springer Texts in Statistics, 2004.

[3] P. Glasserman, *Monte Carlo Methods in Financial Engineering*, Springer Stochastic Modeling and Applied Probability Series, 2003.

[4] C. Z. Mooney, *Monte Carlo Simulation*, Sage, 1997.

5

Network Modeling

This chapter expands upon the concepts introduced in earlier chapters and applies them to the area of network modeling and simulation. We discuss the different applications of modeling and simulation in the design and optimization of networked environments. We introduce the network modeling project life cycle, and expose the reader to some of the particular considerations when modeling network infrastructures. Finally, the chapter attempts to describe applications of network modeling within the linkage between network modeling and business requirements.

Although we will focus on network modeling and simulation in the context of computer communication networks, the content of this chapter applies to networks in general. In the case of computer communication networks, this influence is codified by the transmission of data packets, wireless signals, etc. Historically, the first network simulations were developed in the area of computer communication systems, and thus many of the available tools and frameworks are tailored to this application area. Nevertheless, most network simulation tools are amenable to being applied in a wide range of different problem domains.

In the area of computer network design and optimization, software simulators are a valuable tool given today's complex networks and their protocols, architectures, and dynamic topologies. Designers can test their new ideas and carry out performance-related studies, being freed from the burden of the "trial and error" hardware implementations. A typical *network simulator* can provide the engineer/programmer with the abstraction of multiple threads of control and inter-thread communication. Functions and protocols are described by either a finite-state machine or native programming code, or a combination of both. A simulator typically comes with a set of predefined modules and user-friendly GUI. Some network simulators even provide extensive support for visualization and animation. There are also emulators such as the Network Emulation Tool (NIST Net). By operating at the IP level, it can

Network Modeling and Simulation   M. Guizani, A. Rayes, B. Khan and A. Al-Fuqaha
© 2010 John Wiley & Sons, Ltd.

emulate critical end-to-end performance characteristics imposed by various wide area network situations or by various underlying subnetwork technologies in a laboratory test-bed environment.

5.1 Simulation of Networks

Simulation of a network frequently requires a collection of interdependent nested concurrent sub-simulations arising from:

- The entities that are the nodes of the network, and the processes therein.
- The links that are the edges of the network, and the processes therein.

In the specific case of computer communication networks, these elementary sub-components span functionally distinct elements of the system such as networks, links, circuits, the propagation medium, etc. In practice, simulation of these elementary subcomponents can be conducted using different approaches—with certain approaches being better suited to certain components. A broad classification of approaches would include:

- Discrete-event-driven simulations
- Time-driven simulations
- Algorithm simulations
- Circuit simulations
- Physical media simulations.

Different simulation tools specialize in one or more of these simulation approaches. For example, ns-2 and OPNET deal primarily with event-driven simulations, while SPW, COSSAP, and Simulink/MATLAB focus on time-driven simulations. TI CodeComposer provides a platform for algorithm simulation, while NC-VHDL/ Verilog and Scirroco support circuit simulations. PSpice, ADS, and XFDTD are often used for simulation of radio frequency channels and other media. Within a given system, the constituent subcomponents to be simulated usually exhibit a natural hierarchical structure; for example, simulating a network requires simulating event-driven protocols operating at each node. But simulating a protocol requires simulating a single packet transmission across a wireless link, and this in turn requires modeling how a radio frequency transmission fades in the medium of air. In short, a hierarchy of subcomponents of a system induces a hierarchy of simulations, each of which usually requires a different simulation tool. An example hierarchy for a case of computer communication networks is depicted in Figure 5.1.

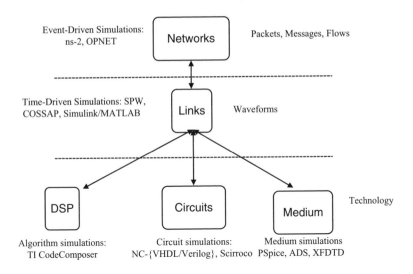

Figure 5.1 Simulation hierarchy

5.2 The Network Modeling and Simulation Process

Figures 5.2 and 5.3 show successive refinements of a block diagram describing the typical phases in the modeling and simulation process. Within these, block 1 represents determination of the actual system to be modeled and simulated. Blocks 2 and 3 constitute *modeling* and will be discussed in Section 5.3. Blocks 4 and 5 constitute *model implementation* in computational terms and instantiation via input/parameter selection. Block 6 represents the actual *simulation execution*, and block 7 represents *analysis of simulation data* generated on performance measures through simulations. Blocks 4–7 are very much dependent on the choice of the simulation package, which, as noted above, should be deliberately selected based on the nature of the subcomponent being simulated. These will be covered in detail in Section 5.4.

Ideally, a mathematical model perfectly describes the evolution of the actual system, but in practice this is computationally not feasible. In addition, it may introduce many parameters that have to be measured in the real world in order to be initialized and used in the simulation. A posteriori to the simulation studies, one can determine which parameters were irrelevant in the sense that they have introduced additional complexity into the model without qualitatively altering the conclusions of the investigation. The question is how to determine this a priori to the simulations. In practice, this depends on the individual's experience and is considered as an art. The general approach taken is to consider evermore complex models, execute simulations, and validate the simulation outcomes with observations of the actual system in an upward

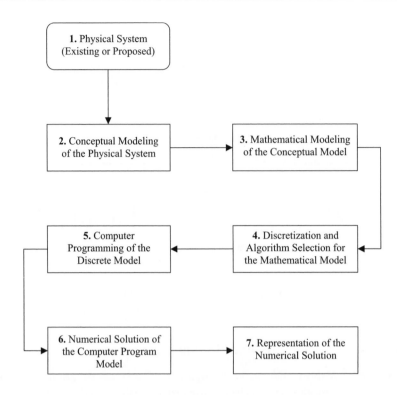

Figure 5.2   Phases in the modeling and simulation process

increasing design complexity. Blocks 2–7 in Figure 5.2 actually form a closed loop, with validation against the physical system playing the role of the return arc. Following this recipe yields a model satisfying Occam's razor—the simplest model which explains the observed phenomenon. A modeling and simulation cycle is shown in Figure 5.4.

5.3 Developing Models

The modeling process (blocks 2 and 3) frequently requires making simplifying assumptions and introducing approximations to reduce the model's complexity. These simplifications can be carried out at two distinct levels:

1. **Modeling:** Here, we simplify the functional description of network elements. For example, we might consider packet transmission to be error free, or we might consider that channel contention due to simultaneous radio frequency transmission is negligible. Such simplifications themselves generally fall into one of three categories:

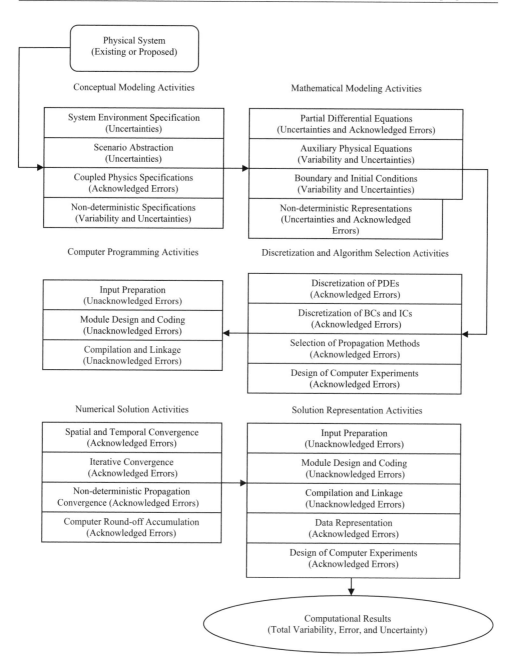

Figure 5.3   Phases in the modeling and simulation process

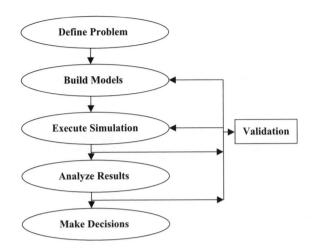

Figure 5.4   Modeling and simulation cycle

(a) *System modeling*: Simplifications at the highest level of description of the interactions *between* distinct elements of the simulation; complexity reduction. As an example, assume that we are simulating a "vehicular network" and we choose to assume that the nodes only move on a Cartesian lattice.

(b) *Element/device modeling*: Simplifications at the level of description of the behavior *within* a single element of the simulation. Assume we are simulating a routing protocol, and we choose to assume that each constituent switch is either fully functional or else fails completely (i.e., there are no partial failures). Another example is the simulation of the propagation of a social network, where we assume that each individual adopts the same probability whenever any item of hardware in the social network is expressed by any of its peers (regardless of the specific identity of the peer).

(c) *System models and element/device models*: These are frequently expressed with reference to random processes. For example, a network of computers might be connected using links that are drawn uniformly at random from the set of all distinct pairs of nodes. An individual element in a social network might have traits selected at random from the normal distributions centered at the mean value observed in the actual population census. Frequently, simplifications of system and device/element models entail simplifications in the distributions referenced by their constituent random processes. This brings us to a third type of simplification called *random process modeling*.

2. **Performance Evaluation:** Here, we simplify the measurements being made of the simulation to provide less precise but potentially more useful estimates of the system's behavior.

5.4 Network Simulation Packages

As noted earlier, blocks 4–7 are very much dependent on the choice of simulation tools. Some examples of academic simulators include the following:

- **REAL** is a simulator for studying the dynamic behavior of flow and congestion control schemes in packet switched data networks. Network topology, protocols, data, and control parameters are represented by Scenarios, which are described using NetLanguage, a simple ASCII representation of the network. About 30 modules are provided which can exactly emulate the actions of several well-known flow control protocols.
- **INSANE** is a network simulator designed to test various IP-over-ATM algorithms with realistic traffic loads derived from empirical traffic measurements. Its ATM protocol stack provides real-time guarantees to ATM virtual circuits by using Rate Controlled Static Priority (RCSP) queuing. A protocol similar to the Real-Time Channel Administration Protocol (RCAP) is implemented for ATM signaling. A graphical simulation monitor can provide an easy way to check the progress of multiple running simulation processes.
- **NetSim** is intended to offer a very detailed simulation of Ethernet, including realistic modeling of signal propagation, the effect of the relative positions of stations on events in the network, the collision detection and handling process, and the transmission deferral mechanism. However, it cannot be extended to address modern networks.
- **Maisie** is a C-based language for hierarchical simulation, or, more specifically, a language for parallel discrete-event simulation. A logical process is used to model one or more physical processes; the events in the physical system are modeled by message exchanges among the corresponding logical processes in the model. Users can also migrate into recent extensions: Parsec and MOOSE (an object-oriented extension).
- **OPNET** (Optimized Network Engineering Tool) is an object-oriented simulation environment that meets all these requirements and is the most powerful general-purpose network simulator available today. OPNET's comprehensive analysis tool is especially ideal for interpreting and synthesizing output data. A discrete-event simulation of the call and routing signaling was developed using a number of OPNET's unique features such as the dynamic allocation of processes to model virtual circuits transiting through an ATM switch. Moreover, its built-in Proto-C language support gives it the ability to realize almost any function and protocol. OPNET provides a comprehensive development environment for the specification, simulation, and performance analysis of communication networks. A large range of communication systems from a single LAN to global satellite networks can be supported. Discrete-event simulations are

used as the means of analyzing system performance and behavior. Key features of OPNET include:

- *Modeling and simulation cycle*: OPNET provides powerful tools to assist users to go through three of the five phases in a design circle (i.e., the building of models, the execution of a simulation, and the analysis of the output data).
- *Hierarchical modeling*: OPNET employs a hierarchical structure to modeling. Each level of the hierarchy describes different aspects of the complete model being simulated.
- *Specialized in communication networks*: Detailed library models provide support for existing protocols and allow researchers and developers to either modify these existing models or develop new models of their own.
- *Automatic simulation generation*: OPNET models can be compiled into executable code. An executable discrete-event simulation can be debugged or simply executed, resulting in output data. This sophisticated package comes complete with a range of tools that allow developers to specify models in detail, identify the elements of the model of interest, execute the simulation, and analyze the generated output data. A simulation layout is shown in Figure 5.5.
- **SimJava** is a discrete-event, process-oriented simulation package. It is an API that augments Java with building blocks for defining and running simulations. The original SimJava was based on HASE + +, a C + + simulation library, which was in turn based on SIM + +. A SimJava simulation is a collection of entities, each running in its own thread. These entities are connected together by ports and can communicate with each other by sending and receiving event objects. A central system class controls all the threads, advances the simulation time, and delivers the

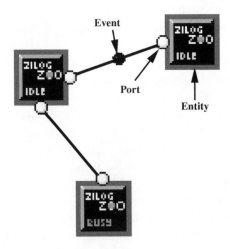

Figure 5.5   A simulation layout

events. The progress of the simulation is recorded through trace messages produced by the entities and saved in a file. Using a programming language to build models (rather than building them graphically) has the advantage that complex regular interconnections are straightforward to specify, which is crucial for some of the networks we are interested in simulating. It also allows the inclusion of existing libraries of code to build simulations. The SimJava package has been designed for simulating fairly static networks of active entities which communicate by sending passive event objects via ports. This model is appropriate for hardware and distributed software systems modeling. As of version 2.0, SimJava has been augmented with considerable statistical and reporting support. The modeler has the ability to add detailed statistical measurements to the simulation's entities and perform output analysis to test the quality of the collected data. Furthermore, much effort has gone into the automation of all possible tasks, allowing the modeler to focus on the pure modeling aspects of the simulation. Automated tasks range from seeding the random number generators used in the simulation to producing detailed and interactive graphs. In the SimJava simulation containing a number of entities, each of which runs in parallel in its own thread, an entity's behavior is encoded in Java using its `body()` method. Entities have access to a small number of simulation primitives:

– `sim_schedule()` sends event objects to other entities via ports;
– `sim_hold()` holds for some simulation time;
– `sim_wait()` waits for an event object to arrive;
– `sim_select()` selects events from the deferred queue.
– `sim_trace()` writes a timestamped message to the trace file.

• **Network simulator (ns-2)** is an object-oriented, discrete-event simulator targeted at research in networking. Developed at UC Berkeley and written in C++ and OTcl, ns is primarily useful for simulating local and wide area networks. It provides substantial support for simulation of TCP, routing, and multicast protocols over wired and wireless (local and satellite) networks. The original ns project began as a variant of the REAL network simulator in 1989 and has evolved substantially over the past few years. The ns-2 simulator covers a large number of protocols, network types, network elements, and traffic models, which are called "simulated objects." One of the main advantages of ns-2 is its Open Source availability.

• **Fast ns-2 simulator:** The fast ns-2 simulator is a modification of the network simulator (ns-2). Fast ns-2 was developed at the Laboratory for Software Technologies at ETH Zurich to enable simulation of large-scale ad hoc wireless networks.[1] Research in ad hoc networks often involves simulators since management and

---

[1] An ad hoc network is formed by wireless mobile nodes (hosts) that operate as terminals as well as routers in the network, without any centralized administration.

operation of a large number of nodes are expensive. However, the widely used original ns-2 did not scale and it was very difficult to simulate even medium-scale networks with 100+ nodes. Hence, ns-2 was modified by ETH Zurich to meet the needs of large ad hoc network simulations. The modified fast ns-2 simulator exploits assumptions in the structure (or absence) of interference in concurrent wireless communication. The modified simulator has simulated populations of up to 3000 nodes so far and works up to 30 times faster than the original version.

- **Simulink:** Simulink is a platform for multi-domain simulation and model-based design for dynamic systems. It provides an interactive graphical environment and a customizable set of block libraries that let one accurately design, simulate, implement, and test control signal processing, communications, and other time-varying systems. It can also be extended for specialized applications. Simulink is an extension to MATLAB, which uses an icon-driven interface for the construction of a block diagram representation of a process. It uses a GUI for solving process simulations. Instead of writing MATLAB code, it simply connects the necessary "icons" together to construct the block diagram. The "icons" represent possible inputs to the system, parts of the systems, or outputs of the system. It allows the user to easily simulate systems of linear and nonlinear ordinary differential equations. Add-on products extend the Simulink environment with tools for specific modeling and design tasks and for code generation, algorithm implementation, test, and verification. Simulink is integrated with MATLAB, providing immediate access to an extensive range of tools for algorithm development, data visualization, data analysis access, and numerical computation.

In the next sections of this chapter we will discuss in more detail the OPNET network simulation tool since it is the most used package in academia and industry, as far as we know (at the time of writing this book).

5.5 OPNET: A Network Simulation Package

In the next chapter, we will be designing our own event-based network simulation framework, from the ground up, on top of the discrete-event framework we developed in Chapter 2. The purpose of this exercise is to understand how such a framework operates "under the hood." Before we embark on this task, it will be illuminating to consider the features offered by a "professional-grade" event-based network simulation tool, as a "gold standard" of comparison. The simulation tool we have in mind is OPNET.

OPNET (Optimized Network Engineering Tools) has achieved widespread use in academia, industry, and government. The US Army adopted OPNET as a standard under the auspices of the Army Enterprise Strategy under the leadership of the

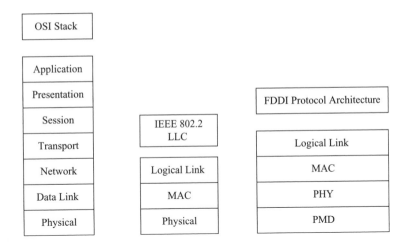

Figure 5.6   Reference models and architectures

US Army Office of the Director of Information Systems for Command, Control, Communications and Computers. OPNET is widely used in universities as well as many parts of the Department of Defense (DOD). OPNET supports visualization through modeling objectives. It may be described as a communications-oriented simulation language. The single most significant aspect of OPNET is that it provides direct access to the source code coupled with an easy-to-use front end.

A generic approach to network modeling can be constructed using the OSI Reference Model as its basis, as shown in Figure 5.6. This approach allows the implementation of different network protocols which are compatible with the OSI layer boundaries. Pedagogically, this approach has limitations. As illustrated in Figure 5.6, any detailed implementation of an Ethernet model will not directly align with the OSI Reference Model. Other protocols such as Fiber Distributed Data Interface (FDDI) also do not perfectly align with the OSI Reference Model.

OPNET models are composed of three primary model layers: the process layer, the node layer, and the network layer. The lowest modeling layer is the process layer. This modeling hierarchy is illustrated in Figure 5.7. The process model in Figure 5.8 shows

Network Models	Networks and Subnetworks
Node Models	Individual Nodes and Stations
Process Models	STD that defines a node

Figure 5.7   OPNET model hierarchy

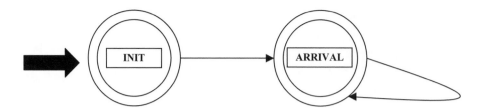

Figure 5.8   State transition diagram in a process model

a state transition diagram (STD) for the generation of packets. Process models are built using finite state-machines (FSMs) described by STDs.

FSMs are an effective means of defining discrete-event systems that maintain state information. FSM-based design provides a means to manage complexity. Complex networks can be broken down into individual states and then each state is defined and implemented.

The source code for each state is readily accessible and modifiable by the user. Each state has *entry execs* and *exit execs*. The term *execs* is used to describe the code executed when a state is entered and when a state is exited. The code defining the state transition between states is also accessible. The next level of abstraction up from the process model is the *node model*. Each element in the node model is either a predefined OPNET artifact or defined by its own STD. Double-clicking on a node model element brings up its underlying process model. Figure 5.9 is an example of a node model that defines a station on an FDDI network. Packets are generated from the source *llc_src*, processed in the *mac* module, and are put on the ring by the *phy_tx* module. Traffic from the ring is received via the *phy_rx* module processed in the *mac* module and finally received and discarded by the *llc_sink* module.

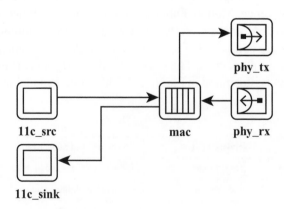

Figure 5.9   Node model (FDDI node)

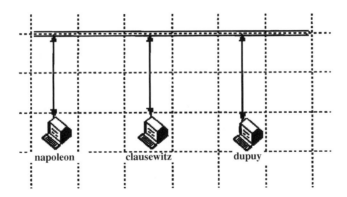

Figure 5.10   Three-node network model

The heart of a node model is either a *processor module* or a *queue module*. Processor modules are used to perform general processing of data packets as specified in the applicable protocol. Queue modules are supersets of processor modules with additional data collection capabilities built in. The *mac* module in Figure 5.9 is an instantiation of a queue module.

The network model is the highest modeling layer in the OPNET model hierarchy. The network model may represent a hierarchy of subnetworks. A network model is shown in Figure 5.10. Each of the stations (nodes) in Figure 5.9 is defined by a node model such as the one in Figure 5.10. Again, each module in a node model is defined by a STD as shown in Figure 5.8 thus conforming to the modeling hierarchy shown in Figure 5.7.

The network model may be used to model a single network, subnet, or segment or a hierarchy of networks, subnetworks, or segments. The segment in Figure 5.10 may be joined with other segments and aggregated into a single subnet icon as shown in Figure 5.11. The operation of a single network segment may now be studied. The implemented functionality of the physical and link layers of the OSI Reference Model are sufficient to model the operation of a single segment. At this point, the individual

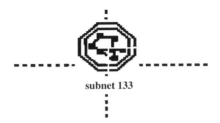

subnet 133

Figure 5.11   Subnetwork of aggregated segments

stations on the segment may be customized if a more detailed representation is desired. Individual workstations or types of workstations may be specially modeled. Special characteristics could be implemented by modifying the individual modules of the station of interest or the physical network line connecting the stations. Many modifications can be made via the built-in menus. However, modifications may be made at the source code level should the menu choices not be fully satisfactory. Adding network services in a TCP/IP network requires the implementation of the Internet Protocol (IP). To simulate the operation of more than one segment, the functionality of network layer services must be added to the model. The node model of Figure 5.10 may be extended by adding modules to implement the IP and Address Resolution Protocol (ARP). ARP tables are implemented statically with entries matching IP addresses to MAC addresses.

5.6 Summary

In this chapter, different applications of modeling and simulation in the design and optimization of networked environments have been discussed. The network modeling project life cycle was introduced with network modeling and simulation processes. Then, different network simulation tools/packages used in academia and industry were introduced, such as REAL, INSANE, NetSim, Maisie, OPNET, SimJava, Network Simulator (ns-2), Fast ns-2 simulator, and Simulink.

Recommended Reading

[1] Online: http://www.opnet.com/
[2] Online: http://www.realsimulation.co.uk/
[3] Online: http://www.boson.com/AboutNetSim.html
[4] Online: http://www.kitchenlab.org/www/bmah/Software/Insane/
[5] Online: http://www.icsa.inf.ed.ac.uk/research/groups/hase/simjava/guide/tutorial.html
[6] Online: http://pcl.cs.ucla.edu/projects/maisie/tutorial/simulation/

6

Designing and Implementing CASiNO: A Network Simulation Framework

In Chapter 2, we designed a framework for discrete-event simulation (FDES) that we used in a case study in Chapter 3. This showed how a FDES can be used to structure a simulation study. Although the framework itself was quite small and consisted of only three classes, `Scheduler`, `SimEnt`, and `Event`, the expressive power of the framework made it easy to structure and reason about simulation design, and implementation. The case study on communities of honeypots in Chapter 3 was an example of a dynamic set of simulation entities that communicate with each other using different types of concrete events in order to advance the state of the system over time. In that study, the pattern of interconnectivity between the simulation entities was dynamic and varied. This is often the case when considering simulations of dynamic networks, since the topology of the network serves only as a relatively loose constraint on the structure of the communication patterns.

Within a single network node, however, the situations that occur are more restricted in scope, as we saw in Chapter 5. Simulations of a single entity/device (i.e. within a single network node) lend themselves to more specialized frameworks that make setting up and conducting intra-device simulation easier. Usually network software (within single node) is organized into layers, with very well defined communication boundaries. Because the interfaces between these "layers" and "modules" are well defined, the implementations can be updated without disturbing the operating system as a whole. This enables abstraction and modular scalable design. In this section, we will define a framework that will allow for modular specification and assembly of

Network Modeling and Simulation   M. Guizani, A. Rayes, B. Khan and A. Al-Fuqaha
© 2010 John Wiley & Sons, Ltd.

dataflow processing modules within a single device. We call this framework the Component Architecture for Simulating Network Objects (CASiNO). The framework described here is a compact Java framework based on an earlier heavy C++ framework of the same name exposited by the authors in [1,2]. The framework we will design and implement in this chapter is an extension of the Framework for Discrete Event Simulation (FDES) described in Chapter 2. Our purpose is to make the general-purpose FDES framework more specialized and better suited for building distributed protocol simulations.

6.1 Overview

The CASiNO framework was originally based on conduits. We feel that it is worth comparing CASiNO to a currently popular network simulator called *ns* [14]. If CASiNO is used to build a network simulator, the resulting simulator and *ns* would indeed have some commonalities in their architecture. The counterpart of the *ns* scheduler is the FDES scheduler. Both `Scheduler` classes run by selecting and executing the next earliest event scheduled. One of the differences between the *ns* and the CASiNO codes is that *ns*, while written in Java, has an "OTcl" interpreter as a front end. Nevertheless, *ns* supports a class hierarchy in C++ and a similar class hierarchy within the OTcl interpreter, thus allowing split-level programming. Another difference is that the CASiNO framework is designed for actually implementing a network protocol as an object-oriented software simulation. CASiNO itself is not a simulator but an object-oriented framework for developing network protocols and simulations.

An object-oriented application is a set of objects working together to achieve a common goal. These objects communicate by sending messages to one another, i.e., by invoking methods of one another. In networking software, one can identify two categories of objects: those that form the architecture of the different protocols; and those that form the packets being sent from entity to entity. We refer to the architecture objects as *conduits*, and the different packets as *visitors*. CASiNO provides `Conduit` and `Visitor` classes; software design using the dataflow architecture centers around the flow of `Visitor` objects through `Conduit` objects. The term *dataflow* here is used to mean coarse-grained "packet" flow, rather than classical fine-grained streams.

A `Conduit` object implements a logical component within a network protocol. Conduits are connected together in a combinatorial graph to form what is called the protocol stack. While isolating its private information, a `Conduit` object has definite points of contact with other conduits. These are referred to as "sides" of the conduit. Just as layers in a network protocol stack typically have interfaces with their immediate higher and lower layers, a conduit has two sides—side A and side B—which are references to neighboring conduits. In much the same way that data passes through each layer of the network protocol stack, so too is it within the CASiNO application. A visitor arriving from side A is acted on at a conduit and then typically passed onward out of side B. Conversely, a visitor arriving from side B might be passed on to side A after being acted on in this conduit. Thus, conduits may be viewed as bidirectional channels that act on transient data units called visitors. Just as two adjacent layers of the network protocol stack interface each other, so too can `Conduit` objects be connected together. The programmer can connect instantiated conduits using the `Conduit` class's static functions:

```
public static void join(Half ce1, Half ce2);
```

which joins two "halfs" of two conduits together. To get the top or bottom half of a conduit, one can use the `Conduit` class's method:

```
public Half getHalf(Side e);
```

passing in either `ASIDE`, or `BSIDE`, which are public constants defined in the `Conduit` class:

```
public static final Side ASIDE = new Side();
public static final Side BSIDE = new Side();
```

The architecture of two connected `Conduit` objects is illustrated by drawing a line between them. For example, Figure 6.1 illustrates that the B side of a `Conduit` object named SSCOP is connected to the A side of a `Conduit` object named SAAL. In this example, the SAAL conduit provides convergence functions of the ATM adaptation layer (AAL) in the control plane of the ATM model [11]. The SSCOP conduit

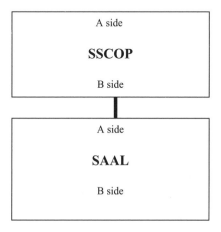

Figure 6.1   The B side of SSCOP connected to the A side of SAAL

implements the Service Specific Connection Oriented Protocol which provides reliable transport for ATM control messages.

Example 6.1 illustrates how a programmer can build such a connected group of conduits.

Example 1: Before connecting the two conduits, they must be created. The following abbreviated Java code instantiates and names them:

```
_sscop = new Conduit("SSCOP", sscop_protocol);
_saal = new Conduit("SAAL", aal_protocol);
```

Then, to connect the B side of SSCOP to the A side of SAAL, the friend functions are used as follows:

```
Conduit.join(_sscop.getHalf(Conduit.BSIDE),
             _saal.getHalf(Conduit.ASIDE));
```

The consequence of the above code is that any visitor leaving out of the A side of conduit SAAL will be passed into the B side of the SSCOP conduit. Likewise, a visitor leaving the SSCOP conduit out of its B side will enter into the SAAL conduit from its A side.

Connecting conduits in this manner determines possible communication patterns between the objects during the application's lifetime. It should be noted, however, that the decision of whether to eject a visitor out of a conduit's A side, or eject it out of the conduit's B side, to absorb the visitor, or delay it, rests in the logic of the actual conduit and visitor. The connectivity of the conduits does not make this decision, rather the geometric configuration of conduit interconnections only limits the range of possible outcomes.

The dataflow architecture is the collection of classes that are used to define the behavior of different modules within the protocol stack that is being implemented. The dataflow architecture is designed to maximize code reuse, make isolation of components easy, and render the interactions between modules as transparent as possible. We will deduce the design of components of CASiNO by considering the needs and characteristics of applications to be developed using CASiNO. Although the focus is on the abstract base classes that form the foundation of CASiNO, from time to time we comment on the necessary requirements placed on programmers in using this library. Whenever possible, we will provide details of CASiNO's internal mechanisms and code. The CASiNO framework is not very large—it consists of about 350 lines of Java code, across 15 classes.

Just as data units are processed in each layer of the network protocol stack (e.g., segmented, assembled, attached to control information, etc.), a visitor's passage through a network of conduits may trigger similar actions. The general manner in which visitors flow through a conduit is referred to as the conduit's *behavior*. Instead of *subclassing* a conduit to specialize in the different behaviors, each conduit *has a* behavior. There are four derived types of behavior: adapter, mux, factory and protocol, based loosely on the concepts presented in [10]. *Behaviors* refer to generic semantic notions: an *adapter* is a starting or ending point of a protocol stack; a *mux* is a one-to-many/many-to-one branch point in a protocol stack; a *protocol* is a FSM; and a *factory* is an object capable of instantiating new conduits and joining them up. Because behaviors represent only general semantic notions, they must be made concrete by composing them with appropriate user-defined actors; each behavior *has an* actor. For example, a protocol behavior requires a user-defined state machine to be concretely defined. Instances of conduits, behaviors, and actors are always in one-to-one-to-one correspondence.

To summarize, CASiNO has three tiers in its design:

1. The conduit level of abstraction represents building blocks of the protocol stack; the function of the `Conduit` class is to provide a uniform interface for joining these blocks together and injecting visitors into the resulting networks.
2. The behavior level of abstraction represents broad classes, each capturing different semantics regarding the conduit's purpose and activity.
3. Finally, at the actor level of abstraction, the user defines concrete implementations. Ultimately, CASiNO application programmers must design a suitable actor for each functionally distinct conduit they intend to instantiate.

As an implementation detail, all conduits, behaviors, and actors must be allocated on the heap. The implication of this to the programmer is that the "new" operator is used for creating an actual object of these classes. Deletion of a conduit automatically deletes any associated `Behavior` and `Actor` objects—the destructor of the conduit eventually deletes them. Because the objects have identical lifetimes, the relation between these objects is more of "aggregation" than "acquaintance" or "association [3, p23]."

There are four subclasses of the `Actor` class: `Terminal`, `Accessor`, `State`, and `Creator`. These subclasses of `Actor` correspond respectively to the subclasses of `Behavior`: `Adapter`, `Mux`, `Protocol`, and `Factory`. There is a relationship among `Conduit`, `Behavior`, and `Actor`. The one-to-one correspondence among the instances of conduits, behaviors, and actors is rooted in the belief that a black-box framework is easier to use, more reusable, and more easily extensible than the "white-box" framework [10]. A "white-box" approach would define the four different types of behaviors as subclass of the `Conduit` class, while the "black-box" approach lets `Conduit` have a reference to these behaviors. A common base class for these four is defined as class `Behavior`, and the `Conduit` class has a reference to a `Behavior` object. This is how a one-to-one correspondence between `Conduit` and `Behavior` is achieved. Regarding the one-to-one correspondence between `Behavior` and `Actor`, the constructor of each subclass of `Behavior` only takes the appropriate corresponding subclass of the `Actor` object as a parameter. At the conduit level of abstraction, the basic unit (conduit) is simply an object that can transport visitors. At the behavior level of abstraction, we identify semantic classes of conduit operation. For example, a mux operates by multiplexing visitors in one direction and demultiplexing them in another direction. A terminal is a source/sink of visitors. A factory is a place where new conduits are made. To fully specify a concrete behavior several details must be specified: for example, for a mux, "how does a mux determine where to demultiplex a visitor to?"; for a factory, "what kind of conduit should be made?"; and so on. This specification process could have been carried out in two ways: (1) by making certain methods abstract in the behavior hierarchy and then requiring concrete versions of the behaviors to implement these abstract methods; or (2) by composition of the specification into the behavior. We chose the latter approach, and we call the specification objects *actors*. If we consider CASiNO as a high-level language, the set of behaviors is in essence the syntactic classes of the language, whereas the set of actors consists of the concrete words within the syntactic classes. We wanted to make the act of extending syntactic classes different from the action of implementing new concrete words within a syntactic class. The latter is commonplace in designing a CASiNO application; the former constitutes extending the CASiNO framework. Conduits receive and are given visitors through their `acceptVisitorFrom()` method. The code below, for example, sends the visitor *v* to the conduit *C*:

```
Visitor v = new PacketVisitor(0, 111);
Conduit C = new Conduit("MyConduit", _my_protocol);
C.acceptVisitorFrom(v, Conduit.ASIDE);
```

Visitors may be thought of as transient objects, usually representing communication messages or dataflow. However, because visitors are full Java objects they are much more than just passive data packets or "buffers.". A visitor is a smart object, with the potential of having its own logic and interacting with the logic of user-defined actors. So at each conduit, a dialog between the visitor and the underlying actor ensues, and the outcome of this dialog determines the side effects within the state of the visitor and the state of the actor. Precisely, the visitor is first informed of the conduit where it is about to find itself situated. This is done by calling the visitor's `arrivedAt()` method:

```
void arrivedAt(Conduit.Half ce);
```

This method may optionally be redefined within the application's derived visitors; in the default implementation, the visitor merely adds to its internal log of conduits it has visited. This log can later be revealed, analyzed, and if necessary cleared. Programmers who use CASiNO must derive all of their visitors from the abstract base class `Visitor`. All visitors must be allocated on the heap. Destruction of a visitor takes place by garbage collection when no reachable references exist:

1. When making a conduit, the programmer is required to specify its behavior. As mentioned earlier, there are four types of behaviors, each representing general semantics of a conduit's purpose and activity:
   (a) **Protocol behavior:** Contains the logic for a FSM.
   (b) **Adapter behavior:** Absorbs and/or generates new visitors; interfaces with non-framework software.
   (c) **Mux behavior:** Demultiplexes visitors to (and multiplexes visitors from) a set of adjacent conduits.
   (d) **Factory behavior:** Creates and attaches new conduits to neighboring muxes.
2. The programmer must fully specify a behavior at the time of construction with an appropriate concrete `Actor` object. There are, correspondingly, four types of actors:
   (a) **State actor:** Required to specify a protocol.
   (b) **Terminal actor:** Required to specify an adapter.
   (c) **Accessor actor:** Required to specify a mux.
   (d) **Creator actor:** Required to specify a factory.

Now that we have a rough outline of how we see the framework operating, let us proceed by defining the `Conduit` and `Visitor` classes. Once that is done, we will proceed with the `Behavior` classes and the corresponding `Actor` interfaces.

6.2 Conduits

Closely associated with a conduit is the notion of a "side." Every conduit has two sides, which we refer to as the `ASIDE` ("A" for above) and the `BSIDE` ("B" for below). Thus, inside the `Conduit` class:

```
public final class Conduit {
}
```

we will have a static class called `Side`:

```
public static class Side {};
public static final Side ASIDE = new Side();
public static final Side BSIDE = new Side();
```

Given these two final public instances `Conduit.ASIDE` and `Conduit.BSIDE`, we introduce the notion of a conduit's "end." Each conduit has two ends, namely its A end and its B end. A conduit's end is represented by an instance of the class `Half`:

```
public static class Half {

  private Conduit _c;
  private Side _e;

  Half(Conduit c, Side e) {
    _c = c;
    _e = e;
  }

  Conduit getConduit() { return _c; }
  Side getSide() { return _e; }

  public String toString() {
    String s = _c.getName();
    if (_e == Conduit.ASIDE) s+= ".ASIDE";
    else if (_e == Conduit.BSIDE) s+= ".BSIDE";
    return s;
  }
}
```

Getting half a conduit then just amounts to calling the method:

```
public Half getHalf(Side e) {
  return new Half(this, e);
}
```

Because the notion of a conduit's end is always in the context of a `Conduit` instance, we will make the `Half` class static and nested inside the `Conduit` class itself. With the notion of a conduit's end now happily represented by an instance of `Conduit.Half`, we proceed with defining the actual `Conduit` class. Each conduit has two neighbors, the neighbor on its A side and the conduit on its B side. For this we introduce a data member inside the `Conduit` class:

```
private final HashMap _neighbors = new HashMap();
```

`HashMap` maps instances of `Side` to instances of `Conduit.Half`. Typically, the _neighbors map will contain precisely two entries: one `Conduit.Half` corresponding to the key `Conduit.ASIDE`, and one `Conduit.Half` corresponding to the key `Conduit.BSIDE`. These two `Conduit.Half`s represent the sides of adjacent conduits. Clearly, we will need getter and setter methods for this `HashMap`:

```
Half getNeighbor(Side e) {
  return (Half)_neighbors.get(e);
}

public void setNeighbor(Side e, Half ce) {
  unsetNeighbor(e);
  _neighbors.put(e, ce);
}

private void unsetNeighbor(Side e) {
  Half neigh = (Half)_neighbors.get(e);
  _neighbors.remove(e);
  if (neigh != null) neigh.unsetNeighbor();
}
```

The last of these methods points out the need for the `Conduit.Half` class to have an `unsetNeighbor()` method:

```
void unsetNeighbor() {
  getConduit().unsetNeighbor(getSide());
}
```

For example, if we want to find out what conduit is connected to the A side of a conduit C, we would simply call:

```
C.getNeighbor(Conduit.ASIDE);
```

What is returned is not a conduit, but a `Conduit.Half`, since it is not sufficient to say that the A side of C is connected to conduit D—one must indicate if it is the A side or the B side of D. We can see now how two `Conduit.Half` objects can be joined in the implementation of the static method `join()`:

```
public static void join(Half ce1, Half ce2) {
  ce1.getConduit().setNeighbor(ce1.getSide(), ce2);
  ce2.getConduit().setNeighbor(ce2.getSide(), ce1);
}
```

When a conduit is named, we must specify its name and the constituent actor specifying its behavior. This implies that we need data members inside `Conduit` that will maintain this data:

```
private final String _name;
private final Behavior _behavior;
```

The constructor of `Conduit` might look something like this:

```
public Conduit(String name, Actor a) {
  _name = name;
  _behavior = Behavior.create(this, a);
  CASINO.register(this);
}
```

The base `Behavior` class has a static method which will make the mediating `Behavior` instance, based on the type of actor that has been provided. For example, if the actor is an accessor, then the behavior's `create()` method will create a mux, since that is the appropriate mediating behavior.

It may be useful to have all conduits register with a central class (named `CASINO`) so that when the simulation ends, the conduits (more specifically, their constituent actors) can be given an opportunity to clean up their resource allocations and record any information useful to the simulation. We do this in the constructor by calling `CASINO.register()`. Let us also define a `shutdown()` method that will be called when the simulation ends:

```
void shutdown() {
  _behavior.shutdown();
  CASINO.deregister(this);
}
```

The call to the behavior's shutdown method will be made to ripple down to the actor, giving it time to respond to the end of the simulation. We can always get a conduit's name by calling:

```
public String getName() {
  return _name;
}
```

Now let us address the heart of a conduit's existence and activity, i.e., accepting and processing visitors. To this end, we specify the method:

```
public void acceptVisitorFrom(Visitor v, Side e) {
  Half route = _behavior.acceptVisitorFrom(v,e);
  if (route != null) {
    route.acceptVisitor(v);
  }
}
```

The call to the behavior's accept method encompasses the conduit's local response to the arrival of the visitor. The return value of this method call is the place to which the visitor must be routed to next. Notice that the conduit delegates to the behavior the decision about where to send the visitor next. The behavior will in turn delegate to the actor—but the nature of this second delegation depends on the specific type of the behavior. The outcome of this double delegation is a determination of where the visitor should go next. The conduit forwards the visitor to the determined destination route, which is the next Conduit.Half to which the visitor must be sent. This points out the need for the Conduit.Half class having an acceptVisitor() method:

```
void acceptVisitor(Visitor v) {
  v.arrivedAt(this);
  getConduit().acceptVisitorFrom(v, getSide());
}
```

Note that before making the conduit accept the visitor via its acceptVisitor-From() method, we notify the visitor that it has arrived at a specific Conduit. Half. One final utility method will prove useful, and we define it statically in the Conduit class:

```
static Side otherSide(Side e) {
  if (e==ASIDE) return BSIDE;
  else return ASIDE;
}
```

This method may be needed to determine "pass-through" behavior—that is, if a visitor came in through the A side, which side should it go out through "usually?" The other side from where it entered. We have now completed the specification of the Conduit class. Let us turn now to visitors.

6.3 Visitors

We know that visitors are the transient elements that flow through conduits. The decisions about how to respond to visitors and where to route them next are determined by behaviors, and ultimately by the actors specifying them. So far, in `Conduit`, the only time that a visitor's methods are called is in `Conduit. Half`'s `accept()` method, which informs the visitor of its impending arrival prior to handing the visitor off to the conduit. It follows that we must define the `arrivedAt()` method for `Visitor`—but what do we put in it? In the base implementation of `Visitor`, we simply keep a log of the conduit names that the visitor has arrived at. This log can be used later for diagnostics and simulation analysis and visualization. Below then is the complete code for the base `Visitor` class:

```
public class Visitor {
  void arrivedAt(Conduit.Half ce) {
    _log.addLast(ce.toString());
  }

  private final LinkedList _log = new LinkedList();

  public void clearLog() {
    _log.clear();
  }

  public String getLog() {
    String s = "";
      for (Iterator it=_log.iterator(); it.hasNext();) {
        String s1 = (String)it.next();
        s += s1;
      }
      return s;
  }
}
```

The `getLog()` and `clearLog()` methods allow external classes to get the log from a visitor, or to zero-out its contents.

6.4 The Conduit Repository

In defining the `Conduit` class, we noted that we need to have a central repository of conduits so that when the simulation ends, all conduits can be given an opportunity to report their results and return system resources to an appropriate state. This central repository is maintained in the `CASINO` class:

```
public final class CASINO {
  static final HashSet _conduits = new HashSet();
  static void register(Conduit c) {
    _conduits.add(c);
  }

  static void deregister(Conduit c) {
    _conduits.remove(c);
  }

  public static void shutdown() {
    while (_conduits.size() > 0) {
      Iterator it = _conduits.iterator();
      Conduit c = (Conduit)it.next();
      c.shutdown();
    }
  }
}
```

Calling `CASINO.shutdown` at the end of the simulation would result in every conduit having its shutdown method being called, sequentially.

6.5 Behaviors and Actors

Recall that behaviors are made in the conduit's constructor by calling the static `Behavior.create()` method, which is supposed to make the appropriate mediating behavior correspond to the type of actor provided. This naturally suggests that the `Behavior` class should have a method like:

```
static Behavior create(Conduit c, Actor a) {
  Behavior b = null;
  if (a instanceof Terminal) b = new Adapter(c, (Terminal)a);
  if (a instanceof State) b = new Protocol(c, (State)a);
  if (a instanceof Accessor) b = new Mux(c, (Accessor)a);
  if (a instanceof Creator) b = new Factory(c, (Creator)a);
  return b;
}
```

Since it is likely that all behaviors will need a reference to their ambient conduit (so that they can understand the geometry of their relationships with their neighbors), it would be a good idea to have a protected data member:

```
protected Conduit _c;
```

in `Behavior`, and require that this be specified in the constructor:

```
protected Behavior(Conduit c) {
  _c = c;
}
```

Since behaviors cannot be made without reference to a conduit, they can have a notion of name that is inherited from the conduit:

```
public String getName() {
  return _c.getName();
}
```

Every behavior is responsible for responding to the arrival of visitors (by delegation to an actor), as well as determining the next destination conduit for the visitor. To this end, it will be helpful to define the notion of a "default" route for the visitor: by default the visitor will be sent to the conduit which is connected to the "other" side—that is, if it came in from the A side, then it will by default be sent to the conduit that is connected to the B side:

```
public final Conduit.Half defaultRoute(Conduit.Side e) {
  return _c.getNeighbor(Conduit.otherSide(e));
}
```

Finally, there are many different classes of behaviors, and each of these will have a different notion of how to accept a visitor, what to do with it, and how to determine where to send it next. It follows that in the `Behavior` class, the method:

```
abstract Conduit.Half acceptVisitorFrom(Visitor v,
  Conduit.Side e);
```

must be abstract. Similarly, since the actions of each `Behavior` subclass are different, how each of the concrete behaviors "shut down" at the end of the simulation is also likely to be quite different. Accordingly, we define:

```
abstract void shutdown();
```

to be an abstract method. Recall that this method is called from inside the conduit's
`shutdown()` method, which is called from within the `shutdown()` method of the
`CASINO` class. Because these two methods are declared abstract, we must make the
`Behavior` class abstract as well:

```
abstract class Behavior { }
```

This completes the specification of the base abstract `Behavior` class.

6.5.1 Adapter–Terminal

Recall that behaviors are made in the conduit's constructor by calling the static
`Behavior.create()` method, which is supposed to make the appropriate mediating behavior correspond to the type of actor provided. This naturally suggests that the
`Adapter` class should have a constructor like:

```
public final class Adapter extends Behavior {
  private Terminal _term;

  Adapter(Conduit c, Terminal t) {
    super(c);
    _term = t;
    _term.setAdapter(this);
  }
}
```

When an adapter is made, it uses the `setAdapter()` method to inform the
terminal that it will be acting as a mediator with the enclosing conduit. What does an
adapter do when it is asked by a conduit to accept a visitor? It delegates to the terminal
actor, allowing it to process and respond to the visitor. Then it returns null, ensuring that
the conduit will not forward this visitor further:

```
Conduit.Half acceptVisitorFrom(Visitor v, Conduit.Side e) {
  _term.acceptVisitor(v);
  return null;
}
```

What does an adapter do when it is asked to shut down at the end of a simulation? It
merely informs the terminal of the fact:

```
void shutdown() { _term.shutdown(); }
```

Recall that an adapter, being a behavior, is a mediator between a conduit and a concrete actor (the terminal). Let us consider whether there are any services that the adapter might need to provide to a terminal actor. A terminal is the sink or *source* of visitors. If it is to be able to act as a source, then the behavior should provide a means for the terminal to send visitors. To this end, we define:

```
public void inject(Visitor v) {
  _c.getNeighbor(Conduit.ASIDE).acceptVisitor(v);
}
```

From this, we see what is required to specify a terminal. These requirements are placed in a terminal interface:

```
public interface Terminal extends Actor {
  public void shutdown();
  public void setAdapter(Adapter a);
  public void acceptVisitor(Visitor v);
}
```

6.5.2 Mux–Accessor

Since behaviors are made in the conduit's constructor by calling the static `Behavior.create()` method, this naturally suggests that the `Mux` class should have a constructor like:

```
public final class Mux extends Behavior {
  private Accessor _accessor;

  Mux(Conduit c, Accessor a) {
    super(c);
    _accessor = a;
    _accessor.setMux(this);
  }
}
```

When a mux is made, it uses the `setMux()` method to inform the accessor that it will be acting as a mediator with the enclosing conduit. What does a mux do when it is asked by a conduit to accept a visitor? If the visitor has entered the mux from the A side, it delegates to the accessor actor, allowing it to process and route the visitor—this is demultiplexing. If it enters from the B side, it simply passes the visitor to the default route—this is multiplexing. The only exception to the multiplexing logic is two special CASiNO internal visitors which are used to dynamically add/remove conduits from the accessor's set of options. These are `AddToMuxVisitor` and `DelFromMuxVisitor`, which result in calling the accessor's `setNextConduit()` and `delConduit()` methods:

```
Conduit.Half acceptVisitorFrom(Visitor v, Conduit.Side e) {
  Conduit.Half next = null;
  if (e == Conduit.ASIDE) {
    next = _accessor.getNextConduit(v);
  }
  else {
    if (v instanceof AddToMuxVisitor) {
      AddToMuxVisitor a2m = (AddToMuxVisitor)v;
      _accessor.setNextConduit(a2m.getVisitor(),
        a2m.getConduitHalf());
      return null;
    }
    else if (v instanceof DelFromMuxVisitor) {
      DelFromMuxVisitor d4m = (DelFromMuxVisitor)v;
      _accessor.delConduit(d4m.getConduitHalf());
      return null;
    }
  }

  if (next==null) {
    next = defaultRoute(e);
  }
  return next;
}
```

What does an adapter do when it is asked to shut down at the end of a simulation? It merely informs the accessor of the fact:

```
void shutdown() {
  _accessor.shutdown();
}
```

From this we see what is required to specify an accessor. These requirements are placed in an `Accessor` interface:

```
public interface Accessor extends Actor {
  public void shutdown();
  public void setMux(Mux m);
  public Conduit.Half getNextConduit(Visitor v);
  public void setNextConduit(Visitor v, Conduit.Half ce);
  public void delConduit(Conduit.Half ce);
}
```

6.5.3 Protocol–State

Since behaviors are made in the conduit's constructor by calling the static `Behavior.
create()` method, this naturally suggests that the `Protocol` class should have a
constructor like:

```
public final class Protocol extends Behavior {
  private State _state;

  Protocol(Conduit c, State s) {
    super(c);
    _state = s;
    _state.setProtocol(this);
  }
}
```

When a protocol is made, it uses the `setProtocol()` method to inform the state
that it will be acting as a mediator with the enclosing conduit. What does a protocol do
when it is asked by a conduit to accept a visitor? It passes the visitor to the state via its
`handle()` method. If the return value of this call is true, the visitor is passed onward
along the default route. If it is false, the visitor's journey ends here:

```
Conduit.Half acceptVisitorFrom(Visitor v, Conduit.Side e) {
  boolean passthru = _state.handle(v, e);
  if (passthru) return defaultRoute(e);
  else return null;
}
```

What does a protocol do when it is asked to shut down at the end of a simulation? It
merely informs the state of the fact:

```
void shutdown() {
  _state.shutdown();
}
```

Recall that a protocol, being a behavior, is a mediator between a conduit and a concrete actor (the state). Let us consider whether there are any services that the protocol might need to provide to a state actor. A state represents a FSM. If it is to be able to initiate sending visitors (not just responds to their arrival), then it must be given this service through a method of the protocol. To this end, we define:

```
public void sendVisitor(Visitor v, Conduit.Side e) {
  _c.getNeighbor(e).acceptVisitor(v);
}
```

Furthermore, the state may wish to terminate its own existence (e.g., a FSM reaches final state and no longer considers itself necessary). To this end, we provide the following method of the `Protocol` class:

```
public void suicide() {
  _c.getNeighbor(Conduit.ASIDE).acceptVisitor(
    new DelFromMuxVisitor(_c.getHalf(Conduit.ASIDE)));
  _c.getNeighbor(Conduit.BSIDE).acceptVisitor(
    new DelFromMuxVisitor(_c.getHalf(Conduit.BSIDE)));
  _c.shutdown();
}
```

Together, the prior design specification of the protocol tells us what the requirements are for the state actor. These requirements are placed in a `State` interface:

```
public interface State extends Actor {
  public void shutdown();
  public void setProtocol(Protocol p);
  public boolean handle(Visitor v, Conduit.Side e);
}
```

6.5.4 Factory–Creator

Since behaviors are made in the conduit's constructor by calling the static `Behavior.create()` method, this naturally suggests that the `Factory` class should have a constructor like:

```
public class Factory extends Behavior {
  private Creator _creator;

  Factory(Conduit c, Creator x) {
    super(c);
```

```
    _creator = x;
    _creator.setFactory(this);
  }
}
```

When a factory is made, it uses the `setFactory()` method to inform the creator that it will be acting as a mediator with the enclosing conduit. What does a factory do when it is asked by a conduit to accept a visitor? It passes the visitor to the creator, via its `create()` method—this results in the dynamic creation of a new conduit. This newly created conduit is then added to the two muxes that lie above and below the factory, using the visitor to generate the appropriate routing table entries in their accessors. This process of augmenting the muxes is achieved using `AddToMuxVisitors`:

```
Conduit.Half acceptVisitorFrom(Visitor v, Conduit.Side e) {
  Conduit con = _creator.create(v, e);

  _c.getNeighbor(Conduit.ASIDE).acceptVisitor(
    new AddToMuxVisitor(v, con.getHalf(Conduit.ASIDE)));

  con.setNeighbor(Conduit.ASIDE, _c.getNeighbor(Conduit.
    ASIDE));

  _c.getNeighbor(Conduit.BSIDE).acceptVisitor(
    new AddToMuxVisitor(v, con.getHalf(Conduit.BSIDE)));

  con.setNeighbor(Conduit.BSIDE, _c.getNeighbor(Conduit.
    BSIDE));

  return con.getHalf(e);
}
```

What does a factory do when it is asked to shut down at the end of a simulation? It merely informs the creator of the fact:

```
void shutdown() {
  _creator.shutdown();
}
```

Together, the prior design specification of the factory tells us what the requirements are for the creator actor. We place these requirements in a `Creator` interface:

```
public interface Creator extends Actor {
  public void shutdown();
  public void setFactory(Factory f);
  public Conduit create(Visitor v, Conduit.Side ce);
}
```

Since all the actor classes have the `shutdown()` method, it makes sense to refactor that method to the base `Actor` interface:

```
interface Actor {
  public void shutdown();
}
```

This completes the design and implementation of the CASiNO dataflow framework. We will now write some small tutorials to illustrate how the components can be used, before considering larger case studies. The tutorials will be built not only over CASiNO, but also over the discrete-event simulation framework developed earlier.

6.6 Tutorial 1: Terminals

A Timed Dataflow from Source to Sink

The first tutorial with the CASiNO framework involves two adapters connected together. One of the adapters (the "source") generates `PacketVisitors` at periodic intervals, injecting them outward. The other adapter simply receives visitors and maintains a count of what it has received. To implement such a simulation we need to define two concrete `Terminal` classes (to specify the adapter's behavior). These we will call `MySource` and `MySink`, respectively. Given these, the main program will be simple to write:

```
public class Run {
  public static void main(String [ ] args) {
    Conduit sc = new Conduit("source", new MySource());
    Conduit tc = new Conduit("sink", new MySink());

    Conduit.join(sc.getHalf(Conduit.ASIDE),
                 tc.getHalf(Conduit.ASIDE));

    Thread t = new Thread(Scheduler.instance());
    t.start();
```

```
      try { t.join(); }
      catch (Exception e) { }

      CASINO.shutdown();
    }
}
```

Let us now turn to the issue of specifying the two terminals. The source terminal needs to be a `SimEnt`, since it needs to periodically generate visitors. Accordingly, we specify it as a subclass of `SimEnt`, implementing the `Terminal` interface:

```
public class MySource extends SimEnt implements Terminal {

  public MySource() {
    super();
    send(this, new Tick(), 10.0);
  }

  public void shutdown() {
    System.out.println("SUMMARY "+getName()+": sent "+_
      seqnum+" packets, received "+_recv+" packets");
  }

  public String getName() {
    if (_a!=null) return _a.getName();
    else return "MySource (unnamed)";
  }

  private Adapter _a;

  public void setAdapter(Adapter a) {
    _a = a;
  }

  public void acceptVisitor(Visitor v) {
    _recv++;
    System.out.println(""+getName()+" received visitor "+v+
      " at time "+Scheduler.getTime());
  }
```

```
int _recv = 0;
int _seqnum = 0;

public void recv(SimEnt src, Event ev) {
  if ((ev instanceof Tick) && (_a != null)) {
    _seqnum++;
    _a.inject(new PacketVisitor(_seqnum, 0));

    if (_seqnum < 10) {
      send(this, ev, 10.0);
    }
    else {
      Scheduler.instance().stop();
    }
  }
}
}
```

Note that the MySource object makes a Tick event in the constructor and sends it
itself. When that Tick is received, it injects a new PackVisitor into the network,
incrementing the sequence number it uses. It does this 10 times, after which (upon
receiving the 10th Tick) it stops the scheduler, causing the simulation thread to end.

Now let us look at the adapter corresponding to the sink. It requires the specification
of a terminal also. But, because it does not need to have asynchronous time (it is purely
reactive to visitors), it does not need to be derived from the SimEnt base class:

```
public class MySink implements Terminal {

  public MySink() {}

  public void shutdown() {
    System.out.println("SUMMARY "+getName()+": received "+
      _recv+" packets");
  }

  public String getName() {
    if (_a!=null) return _a.getName();
    else return "MySink (unnamed)";
  }

  private Adapter _a;
```

```
public void setAdapter(Adapter a) {
  _a = a;
}
int _recv = 0;

public void acceptVisitor(Visitor v) {
  _recv++;
  System.out.println(""+getName()+" received visitor "+
    v+" at time "+Scheduler.getTime());
}
}
```

Finally, we have `PacketVisitor`. It has a sequence number, data, and a checksum. Although we do not use any of these features in the present tutorial, they will be useful for more complicated examples that we put together in later tutorials.

```
public final class PacketVisitor extends Visitor {
  private int _seq;
  private int _data;
  private int _checksum;

  public PacketVisitor(int seq, int data) {
    _seq = seq;
    _data = data;
    _checksum = calcChecksum();
  }

  public int getSeq() {
    return _seq;
  }

  public int calcChecksum() {
    return (_data % 2);
  }

  public boolean isCorrupt() {
    return (calcChecksum() != _checksum);
  }
```

```
  public void corrupt() {
    _data ^= ((int)(2.0 * Math.random()));
  }

  public String toString() {
    return "Packet["+_seq+"]:"+_data+":("+_checksum+") ";
  }
}
```

Compiling this code:

```
javac -g -d ./classes -classpath ../../classes/casino.jar:
  ../../../fw/classes/fw.jar *.java
```

and running it yields:

```
java -cp ./classes:../../classes/casino.jar:
  ../../../fw/classes/fw.jar casino.ex1.Run
```

```
sink received visitor Packet[1]:0:(0) at time 10.0
sink received visitor Packet[2]:0:(0) at time 20.0
sink received visitor Packet[3]:0:(0) at time 30.0
sink received visitor Packet[4]:0:(0) at time 40.0
sink received visitor Packet[5]:0:(0) at time 50.0
sink received visitor Packet[6]:0:(0) at time 60.0
sink received visitor Packet[7]:0:(0) at time 70.0
sink received visitor Packet[8]:0:(0) at time 80.0
sink received visitor Packet[9]:0:(0) at time 90.0
sink received visitor Packet[10]:0:(0) at time 100.0
SUMMARY source: sent 10 packets, received 0 packets
SUMMARY sink: received 10 packets
```

6.7 Tutorial 2: States

Dataflow through a Stateful Protocol

This tutorial shows the operation of a state actor inside a protocol behavior. The concrete State class that we will define (called MyState) will pass visitors through until it has seen five of them. After that, it will drop all future visitors. This is a very simple example of a stateful protocol in the middle of a dataflow pathway, and serves to illustrate the main concepts of the CASiNO framework:

```
public class MyStateMachine implements State {
  private Protocol _p;

  public MyStateMachine() {
  }

  public void shutdown() {
    System.out.println("SUMMARY "+getName()+
      ": passed "+_passed+" packets, dropped "+_dropped+"
      packets");
  }

  public String getName() {
    if (_p!=null) return _p.getName();
    else return "MyStateMachine (unnamed)";
  }

  public void setProtocol(Protocol p) {
    _p = p;
  }

  private int _counter = 0;
  private int _passed = 0;
  private int _dropped = 0;

  public boolean handle(Visitor v, Conduit.Side e) {
    System.out.print(""+getName()+
      " received visitor "+v+" at time "+Scheduler.
      getTime()+" ... ");

    if (v instanceof PacketVisitor) {
      _counter++;
      if (_counter>5) {
        System.out.println("dropping");
        _dropped++;
        return false;
      }
```

```
    else {
      System.out.println ("passing through");
      _passed++;
      return true;
    }
  }
  else return true;
}
}
```

The new state is used to populate a protocol behavior in a conduit which lies in between the two terminals (a MySource and a MySink). The resulting main() function is only a slight extension of what was written for Tutorial 1:

```
public class Run {

  public static void main (String [ ] args) {
    Conduit sc = new Conduit ("source", new MySource ());
    Conduit mc = new Conduit ("fsm", new MyStateMachine ());
    Conduit tc = new Conduit ("sink", new MySink ());

    Conduit.join (sc.getHalf (Conduit.ASIDE),
                  mc.getHalf (Conduit.ASIDE));
    Conduit.join (mc.getHalf (Conduit.BSIDE),
                  tc.getHalf (Conduit.ASIDE));

    Thread t = new Thread (Scheduler.instance ());
    t.start ();
    try { t.join (); }
    catch (Exception e) { }

    CASINO.shutdown ();
  }
}
```

We can now compile the second example and run it:

```
javac -g -d
    ./classes -classpath
../../classes/casino.jar:../../../fw/classes/fw.jar:
    ../ex1/classes *.java
```

```
java -cp ./classes:../../classes/casino.jar:../../../fw/
   classes/fw.jar:../ex1/classes casino.ex2.Run

fsm received visitor Packet[1]:0:(0) at time 10.0 ... passing
   through
sink received visitor Packet[1]:0:(0) at time 10.0
fsm received visitor Packet[2]:0:(0) at time 20.0 ... passing
   through
sink received visitor Packet[2]:0:(0) at time 20.0
fsm received visitor Packet[3]:0:(0) at time 30.0 ... passing
   through
sink received visitor Packet[3]:0:(0) at time 30.0
fsm received visitor Packet[4]:0:(0) at time 40.0 ... passing
   through
sink received visitor Packet[4]:0:(0) at time 40.0
fsm received visitor Packet[5]:0:(0) at time 50.0 ... passing
   through
sink received visitor Packet[5]:0:(0) at time 50.0
fsm received visitor Packet[6]:0:(0) at time 60.0 ... dropping
fsm received visitor Packet[7]:0:(0) at time 70.0 ... dropping
fsm received visitor Packet[8]:0:(0) at time 80.0 ... dropping
fsm received visitor Packet[9]:0:(0) at time 90.0 ... dropping
fsm received visitor Packet[10]:0:(0) at time 100.0 ...
   dropping
SUMMARY fsm: passed 5 packets, dropped 5 packets
SUMMARY sink: received 5 packets
SUMMARY source: sent 10 packets, received 0 packets
```

6.8 Tutorial 3: Making Visitors

Acknowledging Dataflow through a Stateful Protocol

This tutorial shows the operation of a different state actor inside a protocol behavior. This state actor (called the acknowledger) will dynamically instantiate an acknowledgement (as a specialized visitor of type `AckVisitor`) and send it back in the direction from which it receives a `PacketVisitor`. In other words, the acknowledger will have as its `Handle` method the following code:

```
public boolean handle(Visitor v, Conduit.Side e) {
   System.out.println(""+getName()+" received visitor "+v+"
      at time "+Scheduler.getTime());
```

```
  if (v instanceof PacketVisitor) {
    _acked++;
    System.out.println(""+getName()+" sending ack for "+v+"
      ...");
    _p.sendVisitor(new AckVisitor((PacketVisitor)v), e);
  }
  System.out.println(""+getName()+" passing "+v+"
    through... to "+_p.defaultRoute(e));

  return true;
}
```

This of course requires a data member to be added to the `Acknowledger` class:

```
int _acked = 0;
```

An acknowledgement visitor is a visitor which acknowledges the receipt of a particular `PacketVisitor`. The correspondence is codified by having the sequence number of the `AckVisitor` agree with the sequence number of the `PacketVisitor`:

```
final class AckVisitor extends Visitor {
  private int _seq;

  AckVisitor(PacketVisitor pv) {
    _seq = pv.getSeq();
  }

  public String toString() {
    return "Ack["+_seq+"]";
  }
}
```

The remaining methods of the `Acknowledger` `State` class are "standard":

```
private Protocol _p;

public void shutdown() {
  System.out.println("SUMMARY "+getName()+": acked
    "+_acked+" packets");
}
```

```
public String getName() {
  if (_p!=null) return _p.getName();
  else return "Acknowledger (unnamed)";
}

public void setProtocol(Protocol p) {
  _p = p;
}
```

and the implementation of the constructor is trivial:

```
public Acknowledger() {
}
```

The concrete State class that we will define (called MyState) will pass visitors through until it has seen five of them. After that, it will drop all future visitors. This is a very simple example of a stateful protocol in the middle of a dataflow pathway, and serves to illustrate the main concepts of the CASiNO framework.

The acknowledger is used to populate a protocol behavior in a conduit which lies in between the protocol running MyStateMachine and the adapter running the MySink terminal. The resulting main() function is only a slight extension of what was written for Tutorial 2:

```
public class Run {
  public static void main(String [ ] args) {
    Conduit sc = new Conduit("source", new MySource());
    Conduit mc = new Conduit("fsm", new MyStateMachine());
    Conduit ac = new Conduit("acker", new Acknowledger());
    Conduit tc = new Conduit("sink", new MySink());

  Conduit.join(sc.getHalf(Conduit.ASIDE),
               mc.getHalf(Conduit.ASIDE));
  Conduit.join(mc.getHalf(Conduit.BSIDE),
               ac.getHalf(Conduit.ASIDE));
  Conduit.join(ac.getHalf(Conduit.BSIDE),
               tc.getHalf(Conduit.ASIDE));

  Thread t = new Thread(Scheduler.instance());
  t.start();

  try { t.join(); }
  catch (Exception e) { }
```

```
    CASINO.shutdown();
  }
}
```

We can now compile the third example and run it:

```
javac -g -d ./classes -classpath
  ../../classes/casino.jar:../../../fw/classes/fw.jar:
  ../ex1/classes:
  ../ex2/classes *.java

java -cp
  ./classes:../../classes/casino.jar:../../../fw/
    classes/fw.jar:
  ../ex1/classes:../ex2/classes casino.ex3.Run
```

```
fsm received visitor Packet[1]:0:(0) at time 10.0 ... passing
  through
acker received visitor Packet[1]:0:(0) at time 10.0
acker sending ack for Packet[1]:0:(0) ...
fsm received visitor Ack[1] at time 10.0 ... source received
  visitor Ack[1] at time 10.0
acker passing Packet[1]:0:(0) through... to sink.ASIDE
sink received visitor Packet[1]:0:(0) at time 10.0
fsm received visitor Packet[2]:0:(0) at time 20.0 ... passing
  through
acker received visitor Packet[2]:0:(0) at time 20.0
acker sending ack for Packet[2]:0:(0) ...
fsm received visitor Ack[2] at time 20.0 ... source received
  visitor Ack[2] at time 20.0
acker passing Packet[2]:0:(0) through... to sink.ASIDE
sink received visitor Packet[2]:0:(0) at time 20.0
fsm received visitor Packet[3]:0:(0) at time 30.0 ... passing
  through
acker received visitor Packet[3]:0:(0) at time 30.0
acker sending ack for Packet[3]:0:(0) ...
fsm received visitor Ack[3] at time 30.0 ... source received
  visitor Ack[3] at time 30.0
acker passing Packet[3]:0:(0) through... to sink.ASIDE
sink received visitor Packet[3]:0:(0) at time 30.0
fsm received visitor Packet[4]:0:(0) at time 40.0 ... passing
  through
```

```
acker received visitor Packet[4]:0:(0) at time 40.0
acker sending ack for Packet[4]:0:(0) ...
fsm received visitor Ack[4] at time 40.0 ... source received
   visitor Ack[4] at time 40.0
acker passing Packet[4]:0:(0) through... to sink.ASIDE
sink received visitor Packet[4]:0:(0) at time 40.0
fsm received visitor Packet[5]:0:(0) at time 50.0 ... passing
   through
acker received visitor Packet[5]:0:(0) at time 50.0
acker sending ack for Packet[5]:0:(0) ...
fsm received visitor Ack[5] at time 50.0 ... source received
   visitor Ack[5] at time 50.0
acker passing Packet[5]:0:(0) through... to sink.ASIDE
sink received visitor Packet[5]:0:(0) at time 50.0
fsm received visitor Packet[6]:0:(0) at time 60.0 ...
   dropping
fsm received visitor Packet[7]:0:(0) at time 70.0 ...
   dropping
fsm received visitor Packet[8]:0:(0) at time 80.0 ...
   dropping
fsm received visitor Packet[9]:0:(0) at time 90.0 ...
   dropping
fsm received visitor Packet[10]:0:(0) at time 100.0 ...
   dropping
SUMMARY sink: received 5 packets
SUMMARY source: sent 10 packets, received 5 packets
SUMMARY fsm: passed 5 packets, dropped 5 packets
SUMMARY acker: acked 5 packets
```

6.9 Tutorial 4: Muxes

Routing Visitors to Suitable Conduits

This tutorial shows the operation of an accessor actor inside a mux behavior. This accessor actor (called the checker) will determine whether the `PacketVisitor` is corrupted or not, and route the visitor accordingly to one of two protocols, each running an instance of `MyStateMachine`. In order for the example to be complete, we must create a protocol that randomly corrupts `PacketVisitors` as they transit through it:

```
public class Corrupter implements State {

   private Protocol _p;
```

```
public Corrupter() {
}
public void shutdown() {
  System.out.println("SUMMARY "+getName()+": corrupted "+
    _corrupt+" packets");
}

public String getName() {
  if (_p!=null) return _p.getName();
  else return "Corrupter (unnamed)";
}

public void setProtocol(Protocol p) {
  _p = p;
}

int _corrupt = 0;

public boolean handle(Visitor v, Conduit.Side e) {
  System.out.println("Corrupter received visitor "+
    v+" at time "+Scheduler.getTime());
  if (v instanceof PacketVisitor) {
    PacketVisitor pv = (PacketVisitor)v;
    pv.corrupt();
    if (pv.isCorrupt()) {
      _corrupt++;
    }
  }
  return true;
}
}
```

Now given that the corrupter will lie between the source and the checker mux, we know that packets arriving at the checker may or may not be corrupted. When we write the checker, we will have it keep two references to conduits: the conduit to which good (uncorrupt) packets are to be sent; and the conduit to which bad (corrupt) packets are to be sent. Below is the Checker class:

```
public class Checker implements Accessor {
  private Mux _m;

  public Checker() { }
```

```
public void shutdown() {
  System.out.println("SUMMARY "+getName()+": recvd "+
    _good+" good packets, "+_bad+" bad packets");
}

public String getName() {
  if (_m!=null) return _m.getName();
  else return "Checker (unnamed)";
}

public void setMux(Mux m) {
  _m = m;
}

private Conduit.Half _goodside = null;
private Conduit.Half _badside = null;

public void setNextConduit(Visitor v, Conduit.Half ce) {
  if (v instanceof PacketVisitor) {
    PacketVisitor pv = (PacketVisitor)v;
    if (pv.isCorrupt()) _badside = ce;
    else _goodside = ce;
  }
}

public void delConduit(Conduit.Half ce) {
  if (_goodside==ce) _goodside = null;
  if (_badside==ce) _badside = null;
}

int _good = 0; int _bad = 0;

public Conduit.Half getNextConduit(Visitor v) {
  System.out.println(""+getName()+" received visitor "+v+
    " at time "+Scheduler.getTime());

  if (v instanceof PacketVisitor) {
    PacketVisitor pv = (PacketVisitor)v;
    if (pv.isCorrupt()) {
      _bad++;
      return _badside;
```

```
        }
        else {
          _good++;
          return _goodside;
        }
      }
      return null;
    }
}
```

The dataflow architecture then consists of an adapter (MySource), followed by a protocol (Corrupter) followed by a mux (Checker) which is connected to two protocol instances (Acknowledger) which handle good and bad PacketVisitors respectively and generate AckVisitors:

```
public class Run {

  public static void main(String [ ] args) {
      Conduit sc = new Conduit("source", new MySource());
      Conduit mc = new Conduit("corrupter", new Corrupter());
    Conduit.join(sc.getHalf(Conduit.ASIDE),
                  mc.getHalf(Conduit.ASIDE));

    Checker chk = new Checker();
      Conduit cc = new Conduit("checker", chk);
    Conduit.join(mc.getHalf(Conduit.BSIDE),
                  cc.getHalf(Conduit.ASIDE));

      Conduit ac1 = new Conduit("ackgood", new Acknowledger());
      Conduit tc1 = new Conduit("sinkgood", new MySink());
    Conduit.join(ac1.getHalf(Conduit.BSIDE),
                  tc1.getHalf(Conduit.ASIDE));

    PacketVisitor goodv = new PacketVisitor(0,111);
    chk.setNextConduit(goodv, ac1.getHalf(Conduit.ASIDE));
    ac1.setNeighbor(Conduit.ASIDE, cc.getHalf(Conduit.
      BSIDE));

      Conduit ac2 = new Conduit("ackbad", new Acknowledger());
      Conduit tc2 = new Conduit("sinkbad", new MySink());
```

```
Conduit.join(ac2.getHalf(Conduit.BSIDE),
              tc2.getHalf(Conduit.ASIDE));
ac2.setNeighbor(Conduit.ASIDE, cc.getHalf(Conduit.
   BSIDE));

PacketVisitor badv = new PacketVisitor(0,222);
while (!badv.isCorrupt()) badv.corrupt();
chk.setNextConduit(badv, ac2.getHalf(Conduit.ASIDE));

Thread t = new Thread(Scheduler.instance());
t.start();
try { t.join(); }
catch (Exception e) { }

CASINO.shutdown();
}
```

We can now compile the fourth example and run it:

```
javac -g -d ./classes -classpath ../../classes/casino.
   jar:../../../fw/classes/fw.jar:../ex1/classes:../ex2/
   classes:../ex3/classes *.java

java -cp ./classes:../../classes/casino.jar:../../../fw/
   classes/fw.jar:../ex1/classes:../ex2/classes:../ex3/
   classes casino.ex4.Run
```

```
Corrupter received visitor Packet[1]:0:(0) at time 10.0
checker received visitor Packet[1]:0:(0) at time 10.0
ackgood received visitor Packet[1]:0:(0) at time 10.0
ackgood sending ack for Packet[1]:0:(0) ...
Corrupter received visitor Ack[1] at time 10.0
source received visitor Ack[1] at time 10.0
ackgood passing Packet[1]:0:(0) through... to sinkgood.
   ASIDE
sinkgood received visitor Packet[1]:0:(0) at time 10.0
Corrupter received visitor Packet[2]:0:(0) at time 20.0
checker received visitor Packet[2]:0:(0) at time 20.0
ackgood received visitor Packet[2]:0:(0) at time 20.0
ackgood sending ack for Packet[2]:0:(0) ...
Corrupter received visitor Ack[2] at time 20.0
source received visitor Ack[2] at time 20.0
```

```
ackgood passing Packet[2]:0:(0) through... to sinkgood.
  ASIDE
sinkgood received visitor Packet[2]:0:(0) at time 20.0
Corrupter received visitor Packet[3]:0:(0) at time 30.0
checker received visitor Packet[3]:1:(0) at time 30.0
ackbad received visitor Packet[3]:1:(0) at time 30.0
ackbad sending ack for Packet[3]:1:(0) ...
Corrupter received visitor Ack[3] at time 30.0
source received visitor Ack[3] at time 30.0
ackbad passing Packet[3]:1:(0) through... to sinkbad.
  ASIDE
sinkbad received visitor Packet[3]:1:(0) at time 30.0
Corrupter received visitor Packet[4]:0:(0) at time 40.0
checker received visitor Packet[4]:1:(0) at time 40.0
ackbad received visitor Packet[4]:1:(0) at time 40.0
ackbad sending ack for Packet[4]:1:(0) ...
Corrupter received visitor Ack[4] at time 40.0
source received visitor Ack[4] at time 40.0
ackbad passing Packet[4]:1:(0) through... to sinkbad.
  ASIDE
sinkbad received visitor Packet[4]:1:(0) at time 40.0
Corrupter received visitor Packet[5]:0:(0) at time 50.0
checker received visitor Packet[5]:1:(0) at time 50.0
ackbad received visitor Packet[5]:1:(0) at time 50.0
ackbad sending ack for Packet[5]:1:(0) ...
Corrupter received visitor Ack[5] at time 50.0
source received visitor Ack[5] at time 50.0
ackbad passing Packet[5]:1:(0) through... to sinkbad.
  ASIDE
sinkbad received visitor Packet[5]:1:(0) at time 50.0
Corrupter received visitor Packet[6]:0:(0) at time 60.0
checker received visitor Packet[6]:0:(0) at time 60.0
ackgood received visitor Packet[6]:0:(0) at time 60.0
ackgood sending ack for Packet[6]:0:(0) ...
Corrupter received visitor Ack[6] at time 60.0
source received visitor Ack[6] at time 60.0
ackgood passing Packet[6]:0:(0) through... to sinkgood.
  ASIDE
sinkgood received visitor Packet[6]:0:(0) at time 60.0
Corrupter received visitor Packet[7]:0:(0) at time 70.0
checker received visitor Packet[7]:0:(0) at time 70.0
```

```
ackgood received visitor Packet[7]:0:(0) at time 70.0
ackgood sending ack for Packet[7]:0:(0) ...
Corrupter received visitor Ack[7] at time 70.0
source received visitor Ack[7] at time 70.0
ackgood passing Packet[7]:0:(0) through... to sinkgood.
  ASIDE
sinkgood received visitor Packet[7]:0:(0) at time 70.0
Corrupter received visitor Packet[8]:0:(0) at time 80.0
checker received visitor Packet[8]:1:(0) at time 80.0
ackbad received visitor Packet[8]:1:(0) at time 80.0
ackbad sending ack for Packet[8]:1:(0) ...
Corrupter received visitor Ack[8] at time 80.0
source received visitor Ack[8] at time 80.0
ackbad passing Packet[8]:1:(0) through... to sinkbad.
  ASIDE
sinkbad received visitor Packet[8]:1:(0) at time 80.0
Corrupter received visitor Packet[9]:0:(0) at time 90.0
checker received visitor Packet[9]:0:(0) at time 90.0
ackgood received visitor Packet[9]:0:(0) at time 90.0
ackgood sending ack for Packet[9]:0:(0) ...
Corrupter received visitor Ack[9] at time 90.0
source received visitor Ack[9] at time 90.0
ackgood passing Packet[9]:0:(0) through... to sinkgood.
  ASIDE
sinkgood received visitor Packet[9]:0:(0) at time 90.0
Corrupter received visitor Packet[10]:0:(0) at time 100.0
checker received visitor Packet[10]:1:(0) at time 100.0
ackbad received visitor Packet[10]:1:(0) at time 100.0
ackbad sending ack for Packet[10]:1:(0) ...
Corrupter received visitor Ack[10] at time 100.0
source received visitor Ack[10] at time 100.0
ackbad passing Packet[10]:1:(0) through... to sinkbad.
  ASIDE
sinkbad received visitor Packet[10]:1:(0) at time 100.0
```

SUMMARY corrupter: corrupted 5 packets
SUMMARY ackbad: acked 5 packets
SUMMARY ackgood: acked 5 packets
SUMMARY sinkgood: received 5 packets
SUMMARY source: sent 10 packets, received 10 packets
SUMMARY checker: recvd 5 good packets, 5 bad packets
SUMMARY sinkbad: received 5 packets

6.10 Tutorial 5: Factories

Dynamic Instantiation of Conduits

The previous tutorial showed how a mux could be used to route visitors to one of a set of conduits, based on properties of the visitor. The question that naturally arises is "does the set of destinations (that the mux can route visitors to) have to be statically declared at the outset (as was the case in Tutorial 4) or is there a way to create conduits dynamically?" This is precisely the question which motivates the inclusion of the factory behavior in the CASiNO framework. In this tutorial, we will extend the previous code of Tutorial 4, making an acknowledger protocol and connecting it to the checker mux whenever it is needed—in this example, it will occur the first time a corrupted `PacketVisitor` is seen by the checker, as well as the first time an uncorrupted `PacketVisitor` is seen. Let us see how this happens. We will define a factory whose creator will be a class that is capable of creating `Acknowledger` objects. Note that in the `create()` method, the returned protocol conduit has as its state an instance of class `Acknowledger` whose name is either `ackbad-dyn` or `ackgood-dyn`:

```
class AcknowledgerMaker implements Creator {
  private Factory _f;

  AcknowledgerMaker() { }

  public void shutdown() {
    System.out.println("SUMMARY "+getName()+
                       ": made "+_n+" Acknowledgers");
  }

  public String getName() {
    if (_f!=null) return _f.getName();
    else return "AcknowledgerMaker (unnamed)";
  }

  public void setFactory(Factory f) {
    _f = f;
  }
  int _n = 0;

  public Conduit create(Visitor v, Conduit.Side ce) {
    Conduit c = null;
    if (v instanceof PacketVisitor) {
```

```
      PacketVisitor pv = (PacketVisitor)v;
      System.out.println(""+getName()+
        " making an Acknowledger... and hooking it up");
      if (pv.isCorrupt()) c = new Conduit("ackbad-dyn",
        new Acknowledger());
      else c=new Conduit("ackgood-dyn", new Acknowledger());
      _n++;
    }
    return c;
  }
}
```

The operation of the `AcknowledgerMaker` requires seeing how the various conduits are connected together. A `MySource` adapter is connected to a corrupter protocol which is connected to a "top" checker mux which is connected to an `AcknowledgementMaker` which is connected to a "bottom" checker mux. Finally, the bottom checker mux is connected to a `MySink` adapter:

```
public class Run {

  public static void main(String [ ] args) {
    Conduit sc = new Conduit("source", new MySource());
    Conduit mc = new Conduit("corrupter", new Corrupter());
    Conduit.join(sc.getHalf(Conduit.ASIDE),
             mc.getHalf(Conduit.ASIDE));
    Conduit cc1 = new Conduit("checker-top", new Checker());
    Conduit.join(mc.getHalf(Conduit.BSIDE),
             cc1.getHalf(Conduit.ASIDE));
    Conduit am = new Conduit("ackmaker", new
      AcknowledgerMaker());
    Conduit.join(cc1.getHalf(Conduit.BSIDE),
               am.getHalf(Conduit.ASIDE));
    Conduit cc2 = new Conduit("checker-bottom", new
      Checker());
    Conduit.join(am.getHalf(Conduit.BSIDE),
             cc2.getHalf(Conduit.BSIDE));
    Conduit tc = new Conduit("sink", new MySink());
    Conduit.join(cc2.getHalf(Conduit.ASIDE),
               tc.getHalf(Conduit.ASIDE));
```

```
    Thread t = new Thread(Scheduler.instance());
    t.start();
    try { t.join(); }
    catch (Exception e) { }
    CASINO.shutdown();
  }
}
```

The way the system operates is that if either mux receives a `PacketVisitor` which it is unable to route, it sends the visitor to its `BSIDE`, whence it arrives at the `AcknowledgementMaker` factory. The factory creates a new `Acknowledge-mentMaker` and glues the newly created conduit between the two muxes.

We can now compile the fifth example and run it:

```
javac -g -d ./classes -classpath ../../classes/casino.jar:
    ../../../fw/classes/fw.jar:../ex1/classes:../ex2/
    classes:../ex3/classes:../ex4/classes *.java

java -cp ./classes:../../classes/casino.jar:../../../fw/
    classes/fw.jar:../ex1/classes:../ex2/classes:../ex3/
    classes:../ex4/classes casino.ex5.Run
```

```
Corrupter received visitor Packet[1]:0:(0) at time 10.0
checker-top received visitor Packet[1]:1:(0) at time 10.0
ackmaker making an Acknowledger... and hooking it up
ackbad-dyn received visitor Packet[1]:1:(0) at time 10.0
ackbad-dyn sending ack for Packet[1]:1:(0) ...
Corrupter received visitor Ack[1] at time 10.0
source received visitor Ack[1] at time 10.0
ackbad-dyn passing Packet[1]:1:(0) through... to checker-
    bottom.BSIDE
sink received visitor Packet[1]:1:(0) at time 10.0
Corrupter received visitor Packet[2]:0:(0) at time 20.0
checker-top received visitor Packet[2]:0:(0) at time 20.0
ackmaker making an Acknowledger... and hooking it up
ackgood-dyn received visitor Packet[2]:0:(0) at time 20.0
ackgood-dyn sending ack for Packet[2]:0:(0) ...
Corrupter received visitor Ack[2] at time 20.0
source received visitor Ack[2] at time 20.0
ackgood-dyn passing Packet[2]:0:(0) through... to checker-
    bottom.BSIDE
```

```
sink received visitor Packet[2]:0:(0) at time 20.0
Corrupter received visitor Packet[3]:0:(0) at time 30.0
checker-top received visitor Packet[3]:0:(0) at time 30.0
ackgood-dyn received visitor Packet[3]:0:(0) at time 30.0
ackgood-dyn sending ack for Packet[3]:0:(0) ...
Corrupter received visitor Ack[3] at time 30.0
source received visitor Ack[3] at time 30.0
ackgood-dyn passing Packet[3]:0:(0) through... to checker-
   bottom.BSIDE
sink received visitor Packet[3]:0:(0) at time 30.0
Corrupter received visitor Packet[4]:0:(0) at time 40.0
checker-top received visitor Packet[4]:0:(0) at time 40.0
ackgood-dyn received visitor Packet[4]:0:(0) at time 40.0
ackgood-dyn sending ack for Packet[4]:0:(0) ...
Corrupter received visitor Ack[4] at time 40.0
source received visitor Ack[4] at time 40.0
ackgood-dyn passing Packet[4]:0:(0) through... to checker-
   bottom.BSIDE
sink received visitor Packet[4]:0:(0) at time 40.0
Corrupter received visitor Packet[5]:0:(0) at time 50.0
checker-top received visitor Packet[5]:0:(0) at time 50.0
ackgood-dyn received visitor Packet[5]:0:(0) at time 50.0
ackgood-dyn sending ack for Packet[5]:0:(0) ...
Corrupter received visitor Ack[5] at time 50.0
source received visitor Ack[5] at time 50.0
ackgood-dyn passing Packet[5]:0:(0) through... to checker-
   bottom.BSIDE
sink received visitor Packet[5]:0:(0) at time 50.0
Corrupter received visitor Packet[6]:0:(0) at time 60.0
checker-top received visitor Packet[6]:0:(0) at time 60.0
ackgood-dyn received visitor Packet[6]:0:(0) at time 60.0
ackgood-dyn sending ack for Packet[6]:0:(0) ...
Corrupter received visitor Ack[6] at time 60.0
source received visitor Ack[6] at time 60.0
ackgood-dyn passing Packet[6]:0:(0) through... to checker-
   bottom.BSIDE
sink received visitor Packet[6]:0:(0) at time 60.0
Corrupter received visitor Packet[7]:0:(0) at time 70.0
checker-top received visitor Packet[7]:0:(0) at time 70.0
ackgood-dyn received visitor Packet[7]:0:(0) at time 70.0
ackgood-dyn sending ack for Packet[7]:0:(0) ...
```

```
Corrupter received visitor Ack[7] at time 70.0
source received visitor Ack[7] at time 70.0
ackgood-dyn passing Packet[7]:0:(0) through... to checker-
   bottom.BSIDE
sink received visitor Packet[7]:0:(0) at time 70.0
Corrupter received visitor Packet[8]:0:(0) at time 80.0
checker-top received visitor Packet[8]:0:(0) at time 80.0
ackgood-dyn received visitor Packet[8]:0:(0) at time 80.0
ackgood-dyn sending ack for Packet[8]:0:(0) ...
Corrupter received visitor Ack[8] at time 80.0
source received visitor Ack[8] at time 80.0
ackgood-dyn passing Packet[8]:0:(0) through... to checker-
   bottom.BSIDE
sink received visitor Packet[8]:0:(0) at time 80.0
Corrupter received visitor Packet[9]:0:(0) at time 90.0
checker-top received visitor Packet[9]:0:(0) at time 90.0
ackgood-dyn received visitor Packet[9]:0:(0) at time 90.0
ackgood-dyn sending ack for Packet[9]:0:(0) ...
Corrupter received visitor Ack[9] at time 90.0
source received visitor Ack[9] at time 90.0
ackgood-dyn passing Packet[9]:0:(0) through... to checker-
   bottom.BSIDE
sink received visitor Packet[9]:0:(0) at time 90.0
Corrupter received visitor Packet[10]:0:(0) at time 100.0
checker-top received visitor Packet[10]:0:(0) at time 100.0
ackgood-dyn received visitor Packet[10]:0:(0) at time 100.0
ackgood-dyn sending ack for Packet[10]:0:(0) ...
Corrupter received visitor Ack[10] at time 100.0
source received visitor Ack[10] at time 100.0
ackgood-dyn passing Packet[10]:0:(0) through... to checker-
   bottom.BSIDE
sink received visitor Packet[10]:0:(0) at time 100.0
```
SUMMARY sink: received 10 packets
SUMMARY ackmaker: made 2 Acknowledgers
SUMMARY checker-bottom: recvd 0 good packets, 0 bad packets
SUMMARY source: sent 10 packets, received 10 packets
SUMMARY corrupter: corrupted 1 packets
SUMMARY ackbad-dyn: acked 1 packets
SUMMARY ackgood-dyn: acked 9 packets
SUMMARY checker-top: recvd 9 good packets, 1 bad packets

6.11 Summary

In this chapter, we developed an extension to the discrete-event simulation framework of Chapter 2. This framework, called the Component Architecture for Simulating Network Objects (CASiNO), is based on parallel hierarchies of behaviors and actors, codifying the principal functional classes of components within a network protocol simulation. CASiNO provides us with many desirable features:

1. **Ease of design:** Implementations of protocols "from the ground up" (e.g., in a programming language like Java) require substantial effort to implement common communication constructs, such as timers, message queues, etc. Given a general-purpose language like Java, a programmer can build these in many different ways. A framework, on the other hand, assumes responsibility for all the common low-level aspects of the protocol's implementation, while providing high-level abstractions with which a programmer can quickly specify the logical structure of the protocol. A framework's power comes from the way in which it *reduces* the number of ways in which the programmer can implement a protocol. The key is to carry out this reduction without sacrificing the range of what the programmer can accomplish. Implementations that use a framework are more "uniform" in their design because the number of arbitrary choices is reduced, causing programs to conform to the common structure imposed on them. This structure makes the design and maintenance process simpler.

2. **Ease of description:** By providing common high-level abstractions with which protocols can be quickly implemented, a framework also implicitly provides a succinct language by which to describe and document the design of protocol software.

3. **Ease of maintenance:** Ease of descriptions makes it easier to maintain modular communication protocol software.

4. **Portability:** An ideal framework will hide many system-dependent operations, exposing only the high-level abstractions needed for protocol development. Porting a protocol to a new platform amounts to porting the framework itself to the new platform. This process only needs to be carried out once; subsequently, any protocols implemented using the framework will also operate on the new architecture.

5. **Longevity of code:** Because implementations over a framework are easy to describe and easy to port, the learning curve for implementations is also relatively short, making projects less susceptible to personnel and hardware changes.

6. **Visualization:** Because code is written using high-level abstractions, the invocation of these abstractions can be logged at run time, and subsequently visualized.

Recommended Reading

[1] A. Battou, B. Khan, D. Lee, S. Mountcastle, D. Talmage, and S. Marsh, "CASINO: Component Architecture for Simulating Network Objects," *Software Practice and Experience*, vol. 32, pp. 1099–1128, 2002.

[2] A. Battou, B. Khan, D. Lee, S. Marsh, S. Mountcastle, and D. Talmage, "CASiNO: A Component Architecture for Simulation of Network Objects," *Proceedings of the 1999 Symposium on Performance Evaluation of Computer and Telecommunication Systems, Chicago, Illinois, July 11–15*, Society for Computer Simulation, 1999, pp. 261–272.

[3] E. Gamma, R. Helm, R. Johnson, and J. Vlissides, *Design Patterns: Elements of Object-Oriented Software*, Addison-Wesley, 1995.

[4] L. P. Deutsch, "Design reuse and frameworks in the Smalltalk-80 system," in *Software Reusability, Volume II: Applications and Experience* (T. J. Biggerstaff and A. J. Perlis, eds.), Addison-Wesley, 1989.

[5] R. E. Johnson and B. Foote, "Designing reusable classes," *Journal of Object-Oriented Programming*, vol. 1, pp. 22–35, 1988.

[6] S. Mountcastle, D. Talmage, B. Khan, S. Marsh, A. Battou, and D. Lee, "Introducing SEAN: the signaling entity for ATM networks," *Proceedings of IEEE GLOBECOM'00, San Francisco, California*, IEEE, 2000, pp. 532–537.

[7] A. Battou, B. Khan, S. Marsh, S. Mountcastle, and D. Talmage, "Introducing PRouST: the PNNI routing and simulation toolkit," *Proceedings of the IEEE Workshop on High Performance Switching and Routing, Dallas, Texas*, IEEE, 2001, pp. 335–341.

[8] G. B. Brahim, B. Khan, A. Battou, M. Guizani, and G. Chaudhry, "TRON: The Toolkit for Routing in Optical Networks," in *Proceedings of IEEE GLOBECOM'01, San Antonio, Texas*, IEEE, 2001, pp. 1445–1449.

[9] P. Nikander and A. Karila, "A Java Beans component architecture for cryptographic protocols," *Proceedings of the 7th USENIX Security Symposium, San Antonio, Texas*, 1998.

[10] L. Breslau, D. Estrin, K. Fall, S. Floyd, J. Heidemann, A. Helmy, P. Huang, S. McCanne, K. Varadhan, Y. Xu, and H. Yu, "Advances in network simulation," *IEEE Computer*, pp. 59–67, May 2000.

[11] S. Keshav, *An Engineering Approach to Computer Networking: ATM Networks, the Internet, and the Telephone Network*, Addison–Wesley, 1997.

[12] A. S. Tanenbaum, *Computer Networks*, 3rd Edition, Prentice Hall PTR, 1996.

[13] D. E. Comer, *Internetworking with TCP/IP*, 3rd Edition, Volume 1, Prentice Hall, 1995.

[14] Online: http://www.isi.edu/nsnam/ns/

7

Statistical Distributions and Random Number Generation

Generally, modeling and simulation are developed due to the interest in studying and predicting the behaviors of real systems (and hence verifying the correctness of abstract models). Theoretical analysis of models of systems is typically validated by simulation results, which are considered as an empirical proof of the model's correctness. Simulations thus offer a framework for the study of systems and models.

Simulation engineers use simulators to study the behaviors and correctness of systems by observing the output information generated by a simulator given its input specifications. In such cases, a crucial task in developing simulations is to determine the complete input to the simulator. In order to achieve this, one often must make random choices. For example, in a communication system, node A and node B communicate exactly once every T seconds, where T is determined by choosing a random number in the range of [0, 10]. This is an example of a statistical distribution, which is a formal means of specifying the relative likelihood of different system inputs.

In the rest of this chapter, a set of statistical distributions that could be used in simulation as well as a set of random number generation techniques are discussed. We provide an overview of statistical distributions as well as discrete distributions. Then, continuous distributions and descriptions of some known ones are discussed in detail. Then, an overview of the concepts and issues surrounding the generation of random numbers is presented. A random number generator (RNG) is a process that emits a sequence of numbers that permit the construction of a segment of a sample path of the random process being modeled.

Network Modeling and Simulation   M. Guizani, A. Rayes, B. Khan and A. Al-Fuqaha
© 2010 John Wiley & Sons, Ltd.

7.1 Introduction to Statistical Distributions

A *continuous* random variable is a variable that takes random values over a continuous set of possibilities. A *discrete* random variable is a variable that takes random values over a finite (or countably infinite) set of possibilities. The height of a randomly selected human being is an example of a continuous random variable. The year of birth of a randomly selected human being is an example of a discrete random variable.

7.1.1 Probability Density Functions

Let X be a continuous random variable on the interval $[a, b]$. If the probability that X lies in the interval $[a, b]$ can be expressed by the following function:

$$P(a \leq X \leq b) = \int_a^b f(x)dx$$

then the function $f(x)$ is called *probability density function* (PDF) of the random variable. A PDF satisfies the following conditions:

$$f(x) \geq 0 \text{ for all } x \text{ in } [a, b]$$
$$\int_a^b f(x)dx = 1$$
$$f(x) = 0 \text{ if } x \text{ is not in } [a, b].$$

The function $f(x)$ also has the following properties:

$$\text{For any specific value } x_0, P(X = x_0) = 0 = \int_{x_0}^{x_0} f(x)dx$$

$$P(x \leq X \leq b) = P(x < X \leq b) = P(x \leq X < b) = P(x < X < b).$$

Discrete random variables do not have a PDF. However, it is possible to represent certain discrete random variables using a density of probability, via the Dirac delta function.

7.1.2 Cumulative Density Functions

The cumulative density function (CDF) measures the probability that a random variable X assumes a value less than or equal to a specific value. By convention, if the PDF is denoted by $f(x)$, then the CDF is denoted by $F(x)$. Unlike the PDF, the CDF is able to quantify both discrete and continues random variables.

The CDF for a discrete random variable X is:

$$F(x) = \sum_{\text{all } x_i \leq x} p(x_i).$$

The CDF for a continuous random variable is:

$$F(x) = \int_{-\infty}^{x} f(t)dt.$$

Moreover, the probability of a random variable in a range can be expressed as follows:

$$P(a < X \leq b) = F(a) - F(b) \text{ for all } a < b.$$

The function $F(x)$ has the following properties:

(a) F is an increasing function such that, if $a < b$, then $F(a) \leq F(b)$
(b) $\lim_{x \to +\infty} F(x) = 1$
(c) $\lim_{x \to -\infty} F(x) = 0.$

7.1.3 Joint and Marginal Distributions

Suppose that there are two random variables X and Y, which might have the same or different probability distributions. If X and Y are random variables, then (X, Y) is also a random variable taking values in the Cartesian product of the spaces from which X and Y are drawn. The distribution of the random variable (X, Y) is called the *joint distribution* of X and Y, and is denoted by (X, Y). The distributions of X and Y are each referred to as *marginal distributions*. For example, suppose we have two fair dice, a red one and a green one. The roll of the red die is the variable X_1, while the variable X_2 is used to represent the value of the roll of the green die. We record the output X_1, X_2 after each roll of the pair, and define $X = X_1 + X_2$ and $Y = X_1 - X_2$. The probability distribution of (X, Y) is called a *joint distribution* and the probability distribution of X and Y is called a *marginal distribution*.

7.1.4 Correlation and Covariance

The *covariance* of random variables X and Y measures the relation between X and Y. Intuitively, when X and Y are independent of one another, the covariance gives a score around 0. But if the variables are dependent, their covariance is high. The covariance is computed as:

$$\text{cov}(X, Y) = E[(X - \mu_x)(Y - \mu_y)]$$

such that μ_x is the mean of X and μ_y is the mean of Y.

The *correlation* measures the strength of the linear dependence between X and Y. The values of the correlation function are in the range of -1 and 1. The formula to calculate the correlation is:

$$\rho_{x,y} = \frac{\text{cov}(X, Y)}{\sqrt{\text{var}(X)\text{var}(Y)}} \quad \text{where var is the variance function.}$$

7.1.5 Discrete versus Continuous Distributions

Continuous distributions describe the distribution of continuous random variables. A probability distribution is continuous if its CDF is continuous. For example, we can consider the probability distribution of the time we spend waiting for the bus each morning. If the bus arrives every 20 minutes starting at 6.00 a.m. (until 8.00 a.m.) and if we arrive at the bus stop at a random time in this 2 hour interval, then the probability that we wait a period of time t has a continuous distribution (since t is a continuous variable).

Discrete distributions describe continuous distributions which are zero on all but a finite (or countably infinite) set of values. A probability distribution is discrete if its CDF increases in quantized steps. The Bernoulli and binomial distributions are examples of a discrete distribution. As an example of a discrete distribution, consider a game of chance in which we pick numbers from the set $\{1, 2, 3, 4, 5\}$. The number of possible values in this case is finite. Therefore, the game gives rise to a discrete random variable (the value of the number selected) and the distribution of this random variable is a discrete distribution.

7.2 Discrete Distributions

Discrete distributions describe the statistical distribution of discrete random variables. In the following, we discuss four discrete distributions which are the Bernoulli distribution, the binomial distribution, the geometric distribution, and the Poisson distribution.

7.2.1 Bernoulli Distribution

Consider an experiment consisting of n independent trials. The result of each trial j is success or value 1 with probability p or value 0 for a fail with probability $q = 1 - p$.

So, if X is a random variable from a Bernoulli distribution, we have $P(X = 1) = 1 - P(X = 0) = 1 - q = p$.

The equation of the Bernoulli distribution (its probability mass function) is:

$$p(X = x_j) = \begin{cases} p & x_j = 1, j = 1, 2, 3, \ldots, n \\ 1 - p = q & x_j = 0, j = 1, 2, 3, \ldots, n \\ 0 & \text{otherwise.} \end{cases}$$

The mean and variance of X_j are calculated as follows:

$$E(X_j) = 0.q + 1.p = p$$

$$V(X_j) = p(1 - p).$$

As an example of the Bernoulli distribution, we can consider the famous coin-flipping game. This game consists of flipping a (biased) coin many times and each time considering whether we get heads or tails. Sampling a random variable that is known (or assumed) to have a Bernoulli distribution is easy using Java. The following code achieves this, returning true with probability p and false with probability $1 - p$. The code is built using the Java `Math.random()` function, which chooses random numbers on the unit interval uniformly at random (justification of the correctness of this code will be clarified in Section 7.7):

```
public static boolean bernoulli(double p) {
  return Math.random() < p;
}
```

7.2.2 Binomial Distribution

The binomial distribution is a discrete distribution defined by the number of successes in a sequence of n binomial trials. The random variable X that denotes the number of successes in n Bernoulli trials follows the Binomial distribution, where the probability is provided by the following formula:

$$p(X = x) = \begin{cases} \binom{n}{x} p^x (1 - p)^{n - x} & x = 0, 1, 2, \ldots, n \\ 0 & \text{otherwise} \end{cases}$$

such that

$$\binom{n}{x} = \frac{n!}{x!(n - x)!}.$$

The mean and variance of X are calculated as follows:

$$E(X) = np$$
$$V(X) = npq.$$

As an example of the binomial distribution, consider the example of our coin-flipping game. Using the binomial distribution, we can calculate the probability of having exactly x successes in the game. In practice, when sampling a random variable that is known (or assumed) to have a binomial distribution, a normal distribution is approximated. This will be discussed in more detail in Section 7.3.2.

7.2.3 Geometric Distribution

The geometric distribution is a discrete distribution defined by the probability of getting a "run" of consecutive successes in a Bernoulli experiment, followed by a failure (e.g., a sequence of n heads, followed by a tail). The probability of a random variable X that has a geometric distribution is:

$$p(X = k) = p^{k-1}(1 - p).$$

This is based on a Bernoulli experiment that has p as the probability of success (and $1 - p$ as the probability of failure). The mean and variance of X are calculated as follows:

$$E(X) = \frac{1}{p}$$
$$V(X) = \frac{1-p}{p^2}.$$

As an example of the geometric distribution, consider our game of flipping a coin. Using the geometric distribution, we can calculate the probability of getting a run of x heads followed by a single tail in the game. Again, it is easy to sample a random variable that is known (or assumed) to have a geometric distribution, using Java. The following code achieves this, taking as an argument the probability of success in the underlying Bernoulli experiment and returning an integer, chosen according to the geometric distribution. The code is built using the Java `Math.random()` function, which chooses random numbers on the unit interval uniformly at random (again, justification of the correctness of this code will become clear in Section 7.7):

```java
public static int geometric(double p) {
  return (int) Math.ceil(Math.log(Math.random()) / Math.
    log(1.0 - p));
}
```

7.2.4 Poisson Distribution

The PDF of the Poisson distribution is given by the following formula:

$$p(X = x) = \begin{cases} \dfrac{e^{-\infty}\alpha^x}{x!} & x = 0, 1, \dots \\ 0 & \text{otherwise.} \end{cases}$$

The CDF of the function is

$$F(0 < X \le x) = \sum_{i=0}^{x} \frac{e^{-\alpha}\alpha^i}{i!}.$$

The mean and variance of X are follows:

$$E(X) = \alpha$$
$$V(X) = \alpha.$$

As an example of the Poisson distribution, consider a maintenance person who is beeped each time there is a service call. The number of beeps per hour is a random variable, taken to have a Poisson distribution with mean α. In order to calculate the probability that the maintenance person receives no more than m beeps in an hour, the CDF of the Poisson distribution stated above is used. Again, Java can be used to sample a random variable that is known (or assumed) to have a Poisson distribution. The following code achieves this, taking as an argument the intensity α and returning an integer, chosen according to the Poisson distribution of the specified intensity. The code is built using the Java Math.random() function, which chooses random numbers on the unit interval uniformly at random (justification of the correctness of this code will become clear in Section 7.7 when we discuss random variate generation):

```
public static int poisson(double lambda) {
  int k = 0;
  double p = 1.0;
  double L = Math.exp(-lambda);
  do {
    k++;
    p *= Math.random();
  } while (p >= L);
  return k-1;
}
```

7.3 Continuous Distributions

A continuous random variable is a variable that takes random values over a continuous set of possible values. Continuous random variables can be defined using continuous PDFs, and in this section we describe some of the most common choices for these distributions.

7.3.1 Uniform Distribution

A random variable X is uniformly distributed on the interval $[a, b]$ if its PDF is given by the following function:

$$f(x) = \begin{cases} \dfrac{1}{b-a} & a \le x \le b \\ 0 & \text{otherwise.} \end{cases}$$

In this case, the CDF is given by:

$$F(x) = \begin{cases} 0 & x < a \\ \dfrac{x-a}{b-a} & a \le x < b \\ 1 & x \ge b. \end{cases}$$

The mean and variance of X are calculated as follows:

$$E(X) = \frac{a+b}{2}$$

$$V(X) = \frac{(b-a)^2}{12}.$$

As an example of the uniform distribution, let us consider the random variable X such that X measures the time period for a passenger waiting at a bus stop to get a bus. The bus arrives every 20 minutes starting at 6.00 a.m. and ending at 7.00 a.m. The passenger does not know the schedule but arrives at the bus stop randomly between 6.30 a.m. and 7.00 a.m. The waiting time of the passenger follows a uniform distribution. Again, Java can be used to sample a random variable that is known (or assumed) to have a uniform distribution. The following code achieves this, taking as an argument the interval boundaries a and b, and returning a "double," chosen according to the uniform distribution on this interval. The code is built using the Java `Math.random()` function, which chooses random numbers on the unit interval uniformly at random:

```
public static double uniform (double a, double b) {
  return a + Math.random() * (b-a);
}
```

7.3.2 Gaussian (Normal) Distribution

A random variable X with mean $-\infty < x < +\infty$ and variance $\sigma^2 > 0$ has a normal distribution if its PDF is given by the following function:

$$f(x) = \frac{1}{\sigma\sqrt{2\pi}} \exp\left[-\frac{1}{2}\left(\frac{x-\mu}{\sigma}\right)^2 \right], \quad -\infty < x < +\infty.$$

The CDF is given by:

$$F(x) = \int_{-\infty}^{+\infty} f(x)dx = \varphi\left(\frac{x-\mu}{\sigma}\right)$$

$$\text{and } \Phi(z) = \int_{-\infty}^{z} \frac{1}{\sqrt{2\pi}} e^{-z^2/2}, \quad -\infty < z < +\infty.$$

The mean and variance of X are calculated as follows:

$$E(X) = \mu$$
$$V(X) = \sigma^2.$$

Some of the special properties of the normal distribution are:

a. $\lim_{x \to -\infty} f(x) = \lim_{x \to +\infty} f(x) = 0$; the value of $f(x)$ approaches 0 when x approaches negative or positive infinity.
b. $f(\mu - x) = f(\mu + x)$; the PDF is symmetric about μ.
c. The maximum value of the PDF occurs at $x = \mu$.

The normal distribution is used in many areas due to its properties. It is used for example in statistics for the sampling distributions. For example, the height of the population of 25 year olds in the United States might be represented by a normal distribution centered on its mean, whose variance is computed from the census data. Java can also be used to sample a random variable that is known (or assumed) to have a normal distribution. The following code takes the mean and standard deviation of the distribution as arguments, and returns a "double," chosen according to the normal distribution having the specified mean and standard deviation. The code is built using the Java `Math.random()` function, which chooses random numbers on the unit interval uniformly at random (again, justification for the choice of this code will become clear when we discuss Section 7.7):

```
public static double gaussian(double mean, double stddev) {
   double r, x, y;
   do {
     x = uniform(-1.0, 1.0);
     y = uniform(-1.0, 1.0);
     r = x*x + y*y;
   } while (r >= 1 || r == 0);
   return mean + stddev * (x * Math.sqrt(-2 * Math.log(r) / r));
}
```

7.3.3 Rayleigh Distribution

The Rayleigh distribution is defined with reference to two normally distributed random variables X and Y that are uncorrelated and have equal variance; then the distribution of magnitude R:

$$R = \sqrt{X^2 + Y^2}$$

is said to have a Rayleigh distribution. The PDF of the Rayleigh distribution is given by the following function:

$$f(x/\sigma) = \frac{x \exp\left(\frac{-x^2}{2\sigma^2}\right)}{\sigma^2}$$

where σ is the parameter of the distribution and $x \in [0, \infty)$. In this case, the CDF is given by:

$$F(x) = 1 - \exp\left(\frac{-x^2}{2\sigma^2}\right)$$

The mean and variance of X are computed as follows:

$$E(X) = \sigma\sqrt{\pi/2}$$

$$V(X) = \frac{(4 - \pi)}{2}\sigma^2.$$

An example of the Rayleigh distribution arises when we consider the magnitude of a randomly chosen complex number whose real and imaginary parts are both normally distributed with equal variance.

7.3.4 Exponential Distribution

A random variable X is exponentially distributed on the interval $[a, b]$ with rate λ if its PDF is given by the following function:

$$f(x) = \begin{cases} \lambda e^{-\lambda x} & x > 0 \\ 0 & \text{otherwise.} \end{cases}$$

In this case, the CDF is given by:

$$F(x) = \begin{cases} 0 & x < 0 \\ 1 - e^{-\lambda x} & x \geq 0. \end{cases}$$

The mean and variance of X are calculated as follows:

$$E(X) = \frac{1}{\lambda}$$

$$V(X) = \frac{1}{\lambda^2}.$$

The exponential distribution is mostly used to model inter-arrival intervals when arrivals are completely random and the arrival rate is λ. As an example of the exponential distribution, let us consider the random variable X which measures the life of a computer. The life of the computer is exponentially distributed with catastrophic failure rate $\lambda = 1/3$. The probability that the computer has a life of more than 1000 hours, given that it has survived 2500 hours already, is given by:

$$P(X > 3.5 | x > 2.5) = p(X > 3.5) | p(x > 2.5) = p(X > 1) = e^{-1/3}.$$

Java can also be used to sample a random variable that is known (or assumed) to have an exponential distribution. The following code takes the arrival rate of the distribution as an argument and returns a "double," chosen according to the exponential distribution having the specified arrival rate. The code is built using the Java `Math.random()` function, which chooses random numbers on the unit interval uniformly at random (justification of this code will become clear in the discussion of Section 7.7):

```
public static double exp(double lambda) {
   return -Math.log(1 - Math.random()) / lambda;
}
```

7.3.5 Pareto Distribution

The Pareto distribution is used to model random variables of situations where equilibrium is found in the distribution of the "small" to the large. The distribution is sometimes expressed by the "80–20" rule, e.g., 20% of the population owns 80% of the wealth. For a random variable X having a Pareto distribution, the probability that X is greater than a number x is given by:

$$P(X > x) = \left(\frac{x}{x_m}\right)^{-k}$$

where x_m is the minimum value of the distribution and k is the Pareto index. The PDF of the distribution is given by:

$$f(x) = \frac{kx_m^k}{x^{k+1}}.$$

In this case, the CDF is given by:

$$F(x) = 1 - \left(\frac{x_m}{x}\right)^k.$$

The mean and variance of X are calculated as follows:

$$E(X) = \frac{kx_m}{k-1} \quad \text{for } k > 1$$

$$V(X) = \frac{kx_m^k}{(k-1)^2(k-2)} \quad \text{for } k > 2.$$

Java can be used as well to sample a random variable that is known (or assumed) to have a Pareto distribution. The following code takes the Pareto index as an argument and returns a "double," chosen according to the Pareto distribution with the specified index. The code is built using the Java `Math.random()` function, which chooses random numbers on the unit interval uniformly at random (justification of the correctness will become clear after the discussion of Section 7.7:

```
public static double pareto (double alpha) {
  return Math.pow(1 - Math.random(), -1.0/alpha) - 1.0;
}
```

7.4 Augmenting CASiNO with Random Variate Generators

We put all the prior code snippets together into a single StdRandom class, and then add this class to the CASiNO network simulation framework that we developed in Chapter 6. The code for StdRandom is shown below:

```
public class StdRandom {
  public static double uniform(double a, double b) {
    return a + Math.random() * (b-a);
  }

  public static boolean bernoulli(double p) {
    return Math.random() < p;
  }

  public static double gaussian(double mean, double stddev) {
    double r, x, y;
    do {
      x = uniform(-1.0, 1.0);
      y = uniform(-1.0, 1.0);
      r = x*x + y*y;
    } while (r >= 1 || r == 0);
    return mean + stddev * (x * Math.sqrt(-2 * Math.log(r) / r));
  }

  public static int geometric(double p) {
    return (int) Math.ceil(Math.log(uniform(0.0, 1.0)) /
                           Math.log(1.0 - p));
  }

  public static int poisson(double lambda) {
    int k = 0;
    double p = 1.0;
    double L = Math.exp(-lambda);
```

```
do {
  k++;
  p *= uniform(0.0, 1.0);
} while (p >= L);
return k-1;
}

public static double pareto(double alpha) {
  return Math.pow(1 - uniform(0.0, 1.0), -1.0/alpha) - 1.0;
}

public static double exp(double lambda) {
  return -Math.log(1 - Math.random()) / lambda;
}
}
```

7.5 Random Number Generation

Note that the code in all the prior distributions was built over the `Math.random()` subroutine of Java. This shows that it is possible to generate most distributions given access to a random variable that is uniformly distributed on the unit interval. Since computers operate on finite precision representations of real numbers, this task translates concretely to generating random n-bit integers (i.e., random integers in the range of $0 \ldots n$). In this section, methods for the generation of such uniformly distributed random integers are described, including the linear congruential method, combined linear congruential method, and random number streams. Then, in Section 7.6, tests of randomness, including frequency and correlations tests, will be covered.

7.5.1 Linear Congruential Method

The linear congruential method produces a sequence of integers uniformly distributed between 0 and $m - 1$ following the recursive relationship:

$$X_{i+1} = (aX_i + c)\bmod m, \quad i = 0, 1, 2 \ldots.$$

In this relationship a is called the multiplier, c is called the increment, m is the modulus, and X_0 is the seed. The parameters a, c, m, and X_0 affect drastically the statistical property and the cycle of the numbers. When $c = 0$, the form is known as the multiplicative congruential method.

For example, suppose we want to generate a sequence of random numbers using the parameters $a = 19$, $c = 47$, $m = 100$, and $X_0 = 37$. In order to generate the sequence of integers that follow the method, we use a program for the purpose. The listing below provides the program (written in C# language) to generate the sequence of random numbers:

```
public double[] LinearCongruentialMethod(double a, double
  c, double m, double x0, int size)
  {
    double[] results;
  int i;


  results = new double[size];
  results[0] = x0;

  for (i=1;i<size;i++)
  {
    results[i]=((a*results[i-1])+c) % m;
    x0=results[i];

  }
  //The following loop transforms the numbers to U(0,1)
  for (i = 1; i < size; i++)
  {
    results[i] = results[i]/m;
  }
  return results;
}
```

7.5.2 Combined Linear Congruential

As the complexity of systems to be simulated increases and the computing power increases also, a random number generation with period

$$2^{31} - 1 \approx 2 \times 10^9$$

is used. This is needed since it became difficult to generate enough numbers to simulate complex systems (in this case, we assume 32 is the number of bits to code an integer in a computer). An approach to generate more random numbers is to combine two or more multiplicative congruential generators in such a way that the new generator has a much longer cycle. Later, a combined "linear congruential method" to generate random numbers was proposed. In this method it is assumed that there is k multiplicative congruential generators, where the jth generator has prime modulus m_j and multiplier a_j. These are chosen in such a way that the period is $m_j - 1$. The jth generator produces random numbers $X_{i,j}$ that are approximately uniformly distributed from 0 to $m_1 - 1$. The generator has the following form:

$$X_i = \left(\sum_{j=1}^{k} (-1)^{j-1} X_{i,j} \right) \bmod m_i - 1.$$

The maximum possible period for such a generator is given by:

$$P = \frac{(m_1 - 1)(m_2 - 1) \ldots (m_k - 1)}{2^{k-1}}.$$

7.5.3 Random Number Streams

As mentioned earlier, a linear congruential random number generator is character-ized by the period m, the seed X_0, a multiplier, and an increment. The sequence of numbers generated using this method is repeated after the period m. So, any number in the sequence could be considered as the starting point or the seed of the sequence as the generation is cyclic. The random number stream is a convenient way of referring to the starting point and could take any value in the sequence. For a combined random number generator, if the streams are b values apart, the stream i could be defined as the starting seed $S_i = X_{b(i-1)}$ such that $i = 1, 2, \ldots, P/b$ where P is the size of the sample.

7.6 Frequency and Correlation Tests

The goal of random number generation schema is to produce a sequence of random numbers that simulates the ideal properties. Each number returned must appear to be chosen via the uniform distribution, and any two numbers must appear to have been chosen independently. In order to test if a sequence of random numbers has these proprieties, a set of tests needs to be performed. Tests of whether the distribution of the set of random number is uniform are frequently called *frequency tests*. Tests of whether the generated numbers are independent of one another are called *autocorrelation tests*. In what follows, we describe these two kinds of tests.

We begin with frequency tests; these are used to validate if a new random number generator generates a uniform set of random numbers. The two most popular methods that are available for such tests are the Kolmogorov–Smirnov and chi-square tests.

7.6.1 Kolmogorov–Smirnov Test

The test compares the CDF, namely $F(x)$, of the uniform distribution to the empirical CDF, $S_N(x)$, constructed by taking a sample of N generated random numbers. The definitions of $F(x)$ and $S_N(x)$ are as follows:

$$F(x) = x, \quad 0 \leq x \leq 1$$

$$S_N = \frac{\text{number of } R_1, R_2, \ldots, R_N \text{ which are } \leq x}{N}.$$

The Kolmogorov–Smirnov test is based on the largest absolute deviation between $F(x)$ and $S_N(x)$ —that is, based on the statistic of:

$$D = \max|F(x) - S_N(x)|.$$

If the sample statistic D is smaller than the critical value D_α then it is considered that there is no difference between the sample of the random variable and the uniform distribution. Otherwise, we consider that the sample of the random number is not uniform.

For an example of applying the Kolmogorov–Smirnov test, suppose that we have a set of samples of the random variable {0.05; 0.14; 0.44; 0.81; 0.93}. The process of testing the uniformity of the distribution of the set is as follows:

Step 1: Order the set of random values $R_{(i)}$.
Step 2: Calculate values of D^+ and D^- using the following formula (see Table 7.1):

$$D^+ = \max_{1 \leq i \leq N} \left\{ \frac{i}{N} - R_{(i)} \right\} \text{ and } D^- = \max_{1 \leq i \leq N} \left\{ R_{(i)} - \frac{i-1}{N} \right\}.$$

Step 3: Calculate

$$D = \max(D^+, D^-)$$

Table 7.1 Calculations for Kolmogorov–Smirnov test

$R_{(i)}$	0.05	0.14	0.44	0.81	0.93
i/N	0.20	0.40	0.60	0.8	1.00
$i/N - R_{(i)}$	0.15	0.26	0.16		0.07
$R_{(i)} - (I-1)/N$	0.05		0.04	0.21	0.13

which yields $D = 0.26$. For $\alpha = 0.05$ as the significance value and the number of random values $N = 5$, compare D and D_α. It is observed that $D_\alpha = 0.565 > D = 0.26$. Hence, the set of random values of the example are uniformly distributed.

7.6.2 Chi-Square Test

The chi-square test uses the following sample statistics:

$$\chi_0^2 = \sum_{i=1}^{n} \frac{(O_i - E_i)^2}{E_i}$$

where O_i is the observed number in the ith class, N is the total number of observations, n is the number of classes, and E_i is the expected number in the ith class given by:

$$E_i = \frac{N}{n}.$$

If χ_0^2 is calculated using the above statistics and found to be smaller than the theoretical critical value $\chi_{\alpha,N}^2$ such that α is the confidence level that can be accepted, the hypothesis is that the sample is uniformly distributed. Otherwise, the set of random values is not considered as uniformly distributed.

7.6.3 Autocorrelation Tests

Autocorrelation tests are used to test if a set of random numbers is independent or not. The test requires the computation of the autocorrelation between every m numbers (m is the lag) starting with the ith number. So the autocorrelation between the set of random numbers $\{R_i, R_{i+m}, R_{i+2m}, \ldots, R_{i+(M+1)m}\}$ is ρ_{im} such that N is the total number in the sequence and M is the largest number such that $i + (M+1)m \leq N$. The non-zero autocorrelation implies a lack of independence. The test statistic $Z_0 = \hat{\rho}_{im}/\sigma_{\hat{\rho}_{im}}$ is such that Z is normally distributed with a mean 0 and variance 1 and the formula for the estimators is:

$$\hat{\rho}_{im} = \frac{1}{M+1} \left[\sum_{k=0}^{M} R_{i+km} R_{i+(k+1)m} \right] - 0.25$$

$$\sigma_{\hat{\rho}_{im}} = \frac{\sqrt{13M+7}}{12(M+1)}.$$

The autocorrelation test consists of calculating Z_0 using the empirical distribution of the sample of random numbers and comparing the result to the theoretical value taking into consideration the significance level of the test. If $-Z_{\alpha/2} \leq Z_0 \leq Z_{\alpha/2}$ where α is the level of significance and $Z_{\alpha/2}$ is obtained from the distribution table (found in any statistics book), then it is considered that the set of random variable samples has been generated independently.

7.7 Random Variate Generation

After specifying the distribution of the input random variables (Sections 7.2 and 7.3), understanding how to generate random numbers uniformly at random (Section 7.4), and validating their randomness (Section 7.5), we are now interested in generating samples of the random numbers from a specified (non-uniform) distribution. We have already presented Java codes for generating random samples from each of the discrete and continuous distributions in Sections 7.2 and 7.3, but that code was presented without justification. We now attempt to give the theoretical underpinnings justifying that code.

The different methods of generating random variable samples include the inversion method, the accept–reject method, and the importance sampling method.

7.7.1 Inversion Method

This method could be used to sample random numbers using a wide number of discrete distributions. It is mostly used to generate samples from the exponential, the Weibull, and the triangular distributions. The main steps of the inversion method are:

1. Generate a set of uniform random numbers such that $R \sim U(0,1)$.
2. Choose the CDF, $F(x)$, of the random variable to be sampled.
3. Set $F(X) = R$.
4. Solve the equation $F(X) = R$ for X in terms of R.
5. Generate the random variate X_i from R_i using the function $R = F^{-1}(X)$.

As an example of the use of the inversion method, let us use the exponential distribution. The CDF of the exponential distribution is:

$$F(x) = \begin{cases} 0 & x < 0 \\ 1 - e^{-\lambda x} & x \geq 0. \end{cases}$$

In order to generate random numbers with this distribution, we need to generate a set of random number $\{R_i\}$ using the uniform distribution. Then, we generate numbers X_i

Table 7.2 Generation of exponential random numbers with $\lambda = 1$

i	1	2	3	4	5
R_i	0.1306	0.0422	0.6597	0.7965	0.7696
X_i	0.1400	0.0431	1.078	1.592	1.468

using the fact that the inverse of the CDF for the exponential distribution is:

$$F^{-1}(R_i) = -\frac{1}{\lambda}\log(1 - R_i).$$

One simplification that is commonly used here is to replace $(1 - R_i)$ in the inverse function by R_i since both numbers are uniformly distributed. Table 7.2 provides an example of generating random variables using an exponential distribution with mean 1.

Translating these ideas into a Java code, we get the previously stated Java code which takes the arrival rate of the distribution as an argument and returns a "double," chosen according to the exponential distribution having the specified arrival rate:

```
public static double exp(double lambda) {
  return -Math.log(1 - Math.random()) / lambda;
}
```

7.7.2 Accept–Reject Method

This method is used to generate random numbers using a statistical distribution. In this method some of the generated random variables are accepted and others are rejected based on a selection criterion. In the process of generating random numbers using this method, it is assumed that there is a set of uniform random number R_i such that $R \sim U$ (0, 1). There is also a statistical distribution function $g(x)$ which generates random variable X. Then, a function f is defined, which is chosen in the accept–reject method such that $f(x) \leq c * g(x)$ for all x, where c is a constant greater than 1. The main steps of the accept–reject method are:

1. Generate a set of uniform random numbers R_i.
2. Calculate $\alpha = f(X)/c * g(X)$.
3. Generate the random numbers Y_i using the g function ($Y_i = g(R_i)$).
4. If $R_i < \alpha$ then accept $X_i = Y_i$, otherwise reject Y_i and go to step 3.

7.7.3 Importance Sampling Method

This method is a generalization of the accept–reject method where the condition $g(x)$ is greater than $f(x)$, where $f(x)$ is the PDF of the target distribution and $g(x)$ is the sampling distribution. Like the accept–reject method, some of the generated random numbers in this method are accepted and others are rejected based on a selection function. In order to generate random numbers using this method, it is assumed that there is a set of uniform random numbers R_i such that $R \sim U(0, 1)$. There is also a statistical distribution function $g(x)$ which generates random variables X_i. The PDF function f is defined as a target distribution and takes the weight function $w(x) = f(x)/g(x)$. Then, the maximum weight function W is defined such that:

$$W = \max[w(x)] + \varepsilon, \quad \varepsilon > 0.$$

The main steps of the importance sampling method are:

1. Generate a set of uniform random numbers R_i.
2. Calculate $w(x)$ for all x and calculate W.
3. Apply the inversion method using $G(y)$ to obtain random numbers Y_i.
4. Accept Y_i with probability $P = w(y)/W$.

7.7.4 Generate Random Numbers Using the Normal Distribution

The normal distribution does not have a closed form inverse function. So one cannot use the inverse method to generate random numbers. Many methods have been proposed to generate normally distributed random numbers, such as the inverse Gaussian function of the uniform deviates, Teichroew's approach, and the rejection approach. One can generate normal random variates from a chi-square distribution. The method proceeds as follows. Let U_1 and U_2 be two independent uniform random variables on the interval (0, 1). Consider the random variables X_1 and X_2 such that:

$$X_1 = B \cos \theta$$
$$X_2 = B \sin \theta$$

where θ is uniformly distributed and is symmetric and $B_2 = X_1^2 + X_2^2$ has a chi-square distribution with 2 degrees of freedom, or, similarly, an exponential distribution with $\lambda = 1/2$ (mean 2). The CDF of the exponential distribution is:

$$F(x) = \begin{cases} 0 & x < 0 \\ 1 - e^{-\lambda x} & x \geq 0 \end{cases}$$

and the inverse function of $F(x)$ is:

$$F^{-1}(x) = -\frac{1}{\lambda}\log U.$$

We can then transform the normal variables X_1 and X_2 to obtain the following normal variables:

$$X_1 = (-2\log U_1)^{1/2}\cos(2\pi U_2)$$

$$X_2 = (-2\log U_1)^{1/2}\sin(2\pi U_2).$$

These equations produce normal random numbers on the interval $(0, 1)$. In order to generate random numbers using the normal distribution having mean μ and variance σ one needs to transform the generated values using the following equation:

$$Z_i = \mu + \sigma X_i.$$

The equation for X_1 above is used in the program given below to generate normal random numbers using the normal distribution, and is written using the C# source code:

```
public double[] NormalRandomNumberGeneration
  (double[] UniformRandomNumbers,
   double[] UniformRandomNumbers2)
{
  double[] randomNumbers;
  int i, size;
  size = UniformRandomNumbers.Length;
  randomNumbers = new double[size];

  //transform all the uniform random numbers
  for (i = 1; i < size; i++)
  {
    randomNumbers[i] = Math.Sqrt((-2) * Math.Log(Uniform-
                      RandomNumbers[i])) *
                      Math.Cos (2 * Math.PI * Uniform-
                      RandomNumbers[2]);
  }
  return randomNumbers;
}
```

7.7.5 Generate Random Numbers Using the Rayleigh Distribution

Suppose we have the CDF function G for the Rayleigh distribution. Let us calculate $G^{-1}(x)$:

$$\text{Let } G(X) = R$$

$$\Rightarrow 1 - \exp(-X^2/2\sigma^2) = R$$

$$\Rightarrow \exp(-X^2/2\sigma^2) = 1 - R$$

$$\Rightarrow -X^2/2\sigma^2 = \ln(1 - R)$$

$$\Rightarrow X^2 = -2\sigma^2\ln(1 - R)$$

$$R \in [0, 1] \qquad \ln(1 - R) \leq 0$$

$$\Rightarrow X = \sigma\sqrt{-2\ln(1 - R)}.$$

Now, using the inverse method, we can derive the C# source code (given below) for the generation of random numbers using the Rayleigh distribution:

```csharp
public double[] Rayleigh(double[] UniformRandomNumbers,
  double Sigma)
{
  double[] randomNumbers;
  int i, size;
  size = UniformRandomNumbers.Length;
  randomNumbers = new double[size];
  //transform all the uniform random numbers
  for (i = 1; i < size; i++)
  {
    randomNumbers[i] = Sigma * Math.Sqrt((-2) *
                            Math.Log(1 - UniformRandom-
                                Numbers[i]));
  }
  return randomNumbers;
}
```

7.8 Summary

In this chapter, we have discussed a set of statistical distributions that could be used in simulations as well as a set of random number generation techniques. An overview of statistical distributions was presented first. Then discrete distributions were introduced. Continuous distributions and discrete distributions were then discussed in detail. Subsequently, a random number generator (RNG) was defined and its use in simulations was presented.

Recommended Reading

[1] J. Banks, B. L. Nelson, J. S. Carson, D. M. Nicol, and J. CarsonII, *Discrete-event System Simulation*, 5th Edition, Prentice Hall, 2009.

[2] L. Kleinrock, *Queuing Systems: Computer Applications*, Volume 2, John Wiley & Sons, Inc., 1975.

[3] S. Kotz, N. Balakrishnan, and N. L. Johnson, *Continuous Multivariate Distributions*, 2nd Edition, Volume 2, John Wiley & Sons, Inc., 2000.

[4] N. L. Johnson, S. Kotz, and N. Balakrishnan, *Discrete Multivariate Distributions*, John Wiley & Sons, Inc., 1997.

[5] S. Ross, *A First Course in Probability*, 4th Edition, Macmillan, 1994.

[6] D. S. Moore and G. McCabe, *Introduction to the Practice of Statistics*, 3rd Edition, WH Freeman, 1999.

[7] Online: http://www.ds.unifi.it/VL/VL_EN/dist/dist4.html

8

Network Simulation Elements: A Case Study Using CASiNO

In this chapter, we create some useful network simulation elements to serve as building blocks in the network structures that we will consider in the context of queuing theory in Chapter 9. First, we will design and build a traffic generator which emits packets according to a Poisson distribution. Then, we will design a protocol which processes packets, taking a random amount of time to process each packet (with the processing time being determined according to an exponential distribution). Later, we will see how we can address the possibility of an infinite backlog in such a protocol. Subsequently, we will develop an n-way round-robin multiplexer/demultiplexer. Finally, we address the questions of dynamic instantiation and management of the protocol conduits we developed earlier in the chapter.

8.1 Making a Poisson Source of Packets

In Tutorial 1 for CASiNO at the end of Chapter 6, we saw how to make a source of `PacketVisitors` when we made a concrete conduit with adapter behavior by specifying a terminal called `MySource`. Now let us extend the `MySource` class so as to make it a Poisson source of `PacketVisitors`. Here is the relevant code from `MySource` for the constructor, and the `recv()` method:

```
public MySource() {
  super();
    send(this, new Tick(), 10.0);
}
```

Network Modeling and Simulation   M. Guizani, A. Rayes, B. Khan and A. Al-Fuqaha
© 2010 John Wiley & Sons, Ltd.

```
public void recv(SimEnt src, Event ev) {
  if ((ev instanceof Tick) && (_a != null)) {
    _seqnum++;
    _a.inject(new PacketVisitor(_seqnum, 0));


    if (_seqnum < 10) {
      send(this, ev, 10.0);
    }
    else {
      Scheduler.instance().stop();
    }
  }
}
}
```

The original MySource has two data members of interest:

```
int _recv = 0;
int _seqnum = 0;
```

Now when we modify MySource to make it into a PoissonSource, we can simply extend the class so that it will inherit all the common methods from its parent class's implementation:

```
public class PoissonSource extends MySource {
}
```

We need two new data members inside PoissonSource:

```
private int _max;
private double _lambda;
```

whose initial values will be passed as arguments to the constructor where they will get set:

```
public PoissonSource(int numpackets, double lambda) {
  super();
  _max=numpackets;
  _lambda=lambda;
}
```

The `recv()` method of `PoissonSource` is modified to call the `StdRandom` class's static method `poisson()`, which was described at length in Chapter 7:

```
public void recv(SimEnt src, Event ev) {
  if ((ev instanceof Tick) && (_a != null)) {
    _seqnum++;
    _a.inject(new PacketVisitor(_seqnum, 0));

    if (_seqnum < _max) {
      send(this, ev, StdRandom.poisson(_lambda));
    }
  }
}
```

Note how the receipt of an event causes it to be scheduled for redelivery at a time that is chosen according to the Poisson distribution.

8.2 Making a Protocol for Packet Processing

In Tutorial 3 of Chapter 6, we saw how to make a concrete conduit with protocol behavior by specifying a state called `MyStateMachine`. It did very little. It counted the `PacketVisitors` it received, passing on the first five through and killing all subsequent `PacketVisitors`. Here is the relevant code from `MyStateMachine` for its `handle()` method:

```
public boolean handle(Visitor v, Conduit.Side e) {
  System.out.print(""+getName()+" received visitor "+v+
                    " at time "+Scheduler.getTime()+" ... ");
  if (v instanceof PacketVisitor) {
    _counter++;
    if (_counter>5) {
      System.out.println("dropping");
      _dropped++;
      return false;
    }
    else {
      System.out.println("passing through");
      _passed++;
      return true;
```

```
    }
  }
  else return true;
}
```

Note that this code used the following three data members of the `MyState-Machine` class:

```
private int _counter = 0;
private int _passed = 0;
private int _dropped = 0;
```

One could consider `MyStateMachine` to be a processor of `PacketVisitors` that acts on each packet instantly (in terms of simulation time). Clearly, this is quite an idealized model of a packet processor. Now that we understand more about distributions and sampling random variates (discussed in detail in Chapter 7), we would like to write a packet processor module which does not pass incoming `PacketVisitors` onward instantly. Rather, it requires some (simulation) time to process each `PacketVisitor` before sending it on. The basic idea behind the design of the new "more sophisticated" packet processor is to queue each `PacketVisitor` when it arrives and registers an event to be sent to the packet processor sometime in the future (where this time is determined by sampling a random variate having an exponential distribution). Let us see how to achieve this (using code) for the new concrete `State` class we are writing, which will be named `Exponential-Process`:

```
public boolean handle(Visitor v, Conduit.Side e) {

  if (e==Conduit.ASIDE) {
    _fromA_Q.addLast(v);
    send(this, new ServiceCompleteEvent(_fromA_Q),
              StdRandom.poisson(_lambda));
  }

  if (e==Conduit.BSIDE) {
    _fromB_Q.addLast(v);
    send(this, new ServiceCompleteEvent(_fromB_Q),
              StdRandom.poisson(_lambda));
  }
  return false;
}
```

Note that `PacketVisitors` can arrive from either the A side or the B side, so we need two queues, and we need to have `ServiceCompleteEvent` tell us which queue we should process from. So, we need two new data members inside `ExponentialProcess`:

```
protected LinkedList _fromA_Q = new LinkedList();
protected LinkedList _fromB_Q = new LinkedList();
```

The processing of the queues is signaled by the sending of a `ServiceComplete-Event`, which must now be defined:

```
public class ServiceCompleteEvent implements Event {
  LinkedList _q;
  ServiceCompleteEvent(LinkedList q) {
    _q = q;
  }
  LinkedList getQueue() {
    return _q;
  }
  public void entering(SimEnt locale) {}
}
```

What happens when the `ExponentialProcess` receives a `Service-CompleteEvent`? It removes the next `PacketVisitor` off the appropriate queue (as specified in the event) and then forwards this visitor onward through to the other side of the `ExponentialProcess`. The sequence of queuing, registering an event, receiving the event (later), removing the packet, forwarding the packet onward, together create the illusion of a non-zero processing time for the packet to move through the `ExponentialProcess`. Let us write the code for `recv()` which achieves this:

```
public void recv(SimEnt src, Event ev) {

  if (ev instanceof ServiceCompleteEvent) {
    ServiceCompleteEvent sec = (ServiceCompleteEvent)ev;
    LinkedList q = sec.getQueue();

    Visitor v = (Visitor)q.removeFirst();
    if (q==_fromA_Q) {
      _p.sendVisitor(v, Conduit.BSIDE);
    }
```

```
    else {
      _p.sendVisitor(v, Conduit.ASIDE);
    }
  }
}
```

We will be very interested in the state of queue sizes within the `Exponential-Process`. For example, what was the average queue size? What was the maximum queue size etc.? To keep track of this information, let us augment the `Exponential-Process` to monitor certain data for each queue. For example, we can keep for the queue of packets from the A side:

```
private double lastA_notif_time = 0.0;
private int lastA_notif_value = 0;
private double accumA_value = 0.0;
private int maxA_value = 0;
```

and, similarly, for the queue of packets from the B side:

```
private double lastB_notif_time = 0.0;
private int lastB_notif_value = 0;
private double accumB_value = 0.0;
private int maxB_value = 0;
```

Whenever we add or remove a packet from the queue of packets from the A side, we call:

```
void A_notif(int val, boolean print) {
  if (print) System.out.println("fromA_Q has size: "+val);
  if (val>maxA_value) maxA_value=val;
  accumA_value += (Scheduler.getTime() -
                  lastA_notif_time) * (double)lastA_
                  notif_value;
  lastA_notif_value = val;
  lastA_notif_time = Scheduler.getTime();
}
```

and whenever we add or remove a packet from the queue of packets from the B side, we call:

```
void B_notif(int val, boolean print) {
  if (print) System.out.println("fromB_Q has size: "+val);
```

```
    if (val>maxB_value) maxB_value=val;
    accumB_value += (Scheduler.getTime() -
                    lastB_notif_time) * (double)lastB_
                    notif_value;
    lastB_notif_value = val;
    lastB_notif_time = Scheduler.getTime();
}
```

These methods maintain the local statistical information about the mean and maximum size of the queues (from both the A side and the B side). Then, we need some methods to get these performance measurements for the Exponential-Process instance:

```
public double max_fromA_Qsize() {
    return maxA_value;
}

public double max_fromB_Qsize() {
    return maxB_value;
}

public double mean_fromA_Qsize() {
    A_notif(lastA_notif_value, false);
    return accumA_value/lastA_notif_time;
}

public double mean_fromB_Qsize() {
    B_notif(lastB_notif_value, false);
    return accumB_value/lastB_notif_time;
}
```

8.3 Bounding Protocol Resources

In the previous section, we saw how to make a packet processor as a concrete state specifying a protocol behavior. This new packet processor, the Exponential-Process, used two queues, one for packets flowing in from the A side and one for packets flowing in from the B side. Both of these queues could grow arbitrarily large, however. Clearly, this was a very idealized model. Here we extend this model, making it a more realistic class that bounds the size of these queues. In short, the new class, which we will name ExponentialBoundedProcess, will take an additional argument, the maximum queue size. If at any point, adding a

PacketVisitor to a queue will cause the length of the queue to exceed this upper bound, the PacketVisitor will simply be thrown away. The resulting model more accurately represents the finite resource realities faced by most physical systems. Let us see how to extend the ExponentialProcess in order to define the ExponentialBoundedProcess:

```
Public class ExponentialBoundedProcess extends
  ExponentialProcess {
  private int _droppedfromA = 0;
  private int _droppedfromB = 0;
  private int _b;

  public ExponentialBoundedProcess(double lambda, int b) {
    super(lambda);
    _b = b;
  }

  public boolean handle(Visitor v, Conduit.Side ce) {
    if (ce==Conduit.ASIDE) {
      if (_fromA_Q.size() >= _b) _droppedfromA++;
      else return super.handle(v,ce);
    }
    if (ce==Conduit.BSIDE) {
      if (_fromA_Q.size() >= _b) _droppedfromB++;
      else return super.handle(v,ce);
    }
    return false;
  }
}
```

Note how we keep track of the number of packets that get dropped due to the queue bounds being violated.

8.4 Making a Round-Robin (De)multiplexer

In Tutorial 4 of Chapter 6, we saw how to make a (de)multiplexer object (i.e., a mux) which would route PacketVisitors to one of *two* conduits, based on whether the PacketVisitor's payload was corrupted or not. Now let us extend that code, to make a more sophisticated mux—one that will be able to connect to an arbitrary number of conduits.

Our new mux (which we call RoundRobin) should forward PacketVisitors coming in from its A side in a cyclical manner to each of the conduits that have been connected to its B side. This is the demultiplexing behavior. It should forward PacketVisitors coming in from any conduit on its B side to the one conduit on its A side. This is the multiplexing behavior. In order to do this, we need to specify the accessor which is used to route visitors to and join conduits to the B side. RoundRobin will need two data members to keep track of its neighboring conduits on the B side:

```
private HashMap _port2protocol = new HashMap();
private int _m=0;
```

Here the value of _m will always be the number of neighbors on the B side. To achieve round-robin routing, we need to define the getNextConduit() method, since this is consulted by the mux behavior code to determine where to send visitors which enter from the A side:

```
private int _index = 0;
private int _n = 0;

public Conduit.Half getNextConduit(Visitor v) {
  if (v instanceof PacketVisitor) {
    Conduit.Half ce = (Conduit.Half)
      _port2protocol.get(new Integer(_index));
    _index = (_index+1) % _m;
    _n++;
    return ce;
  }
  else return null;
}
```

Finally, to support dynamically binding conduits to the B side, let us introduce a new kind of visitor called the InitializeVisitor, which will specify the key (or ID) to associate with the newly attached conduit. To attach a conduit to the B side of RoundRobin, we would then call:

```
public void setNextConduit(Visitor v, Conduit.Half ce) {
  int id = ((InitializeVisitor)v)._id;
  _port2protocol.put( new Integer(id), ce);
  _m = _port2protocol.size();
}
```

By passing in a suitable `InitializeVisitor`, together with a conduit that we wish to attach, the `InitializeVisitor` will be queried for its ID, which will be used as the key with which to associate the newly attached conduit. A first implementation of `InitializeVisitor` might be:

```
public final class InitializeVisitor extends Visitor {
  public int _id;

  public InitializeVisitor(int id) {
    _id = id;
  }

  public String toString() {
    return "Initialize["+_id+"]";
  }
}
```

8.5 Dynamically Instantiating Protocols

In Tutorial 5 of Chapter 6, we saw how to use a factory to dynamically create conduits and attach them between two muxes. In the previous section, we designed two round-robin muxes. Now, we will specify a `Creator` class which can be used to create a factory conduit that will dynamically create instances of `ExponentialProcess` (and `ExponentialBoundedProcess`) that we designed in Sections 8.2 and 8.3. The only really relevant code from our concrete `Creator` class (which we call `ExpMaker`) is its `create()` method. It will need to know what kind of parameters to pass to the constructors of `ExponentialProcess` and `Exponential-BoundedProcess`, so these values will have to be added to the definition of the `InitializeVisitor` class. Below is the `create()` method of the `ExpMaker` class:

```
int _n = 0;

public Conduit create(Visitor v, Conduit.Side ce) {
  Conduit c = null;
  if (v instanceof InitializeVisitor) {
    InitializeVisitor iv = (InitializeVisitor)v;
    if (iv._bounded) {
      System.out.println(""+getName()+" making EBP,
        hooking it up");
```

```
          c = new Conduit("bexp-"+iv._id,
                     new ExponentialBoundedProcess(iv._lambda,
                     iv._b));
      }
      else {
        System.out.println(""+getName()+" making EP,
          hooking it up");
        c = new Conduit("exp-"+iv._id,
                     new ExponentialProcess(iv._lambda));
      }
      _n++;
    }
    return c;
  }
}
```

We see that when an `InitializeVisitor` arrives, it is queried to determine
if `_bounded` is true or false and to determine if an `ExponentialProcess` or an
`ExponentialBoundedProcess` should be made. The arguments for the
constructors of these two classes are also obtained from `InitializeVisitor`,
which implies that the latter class needs to be augmented:

```
public final class InitializeVisitor extends Visitor {
  public int _id;
  public double _lambda;
  public boolean _bounded;
  public int _b;

  public InitializeVisitor(int id, double lambda,
                                  boolean bounded, int bound) {
    _id = id;
    _lambda = lambda;
    _bounded = bounded;
    _b = bound;
  }

  public String toString() {
    return "Initialize["+_id+"]";
  }
}
```

8.6 Putting it All Together

Let us now build a simple simulation. The system to be simulated consists of a PoissonSource which acts as a source of packets, connected to a RoundRobin mux, which is connected to an ExpMaker factory which is connected to a Round-Robin mux which is connected to a MySink terminal. The system is illustrated in Figure 8.1.

The code that creates this architecture is:

```
Conduit sc = new Conduit("poisson source",
                             new PoissonSource(100, 2.0));
Conduit rc1 = new Conduit("roundrobin-top", new RoundRobin());
Conduit fc = new Conduit("exp-factory", new ExpMaker());
Conduit rc2 = new Conduit("roundrobin-bottom",
  new RoundRobin());
Conduit tc = new Conduit("sink", new MySink());

Conduit.join(sc.getHalf(Conduit.ASIDE), rc1.getHalf
  (Conduit.ASIDE));
Conduit.join(rc1.getHalf(Conduit.BSIDE), fc.getHalf
  (Conduit.ASIDE));
Conduit.join(fc.getHalf(Conduit.BSIDE), rc2.getHalf
  (Conduit.BSIDE));
Conduit.join(rc2.getHalf(Conduit.ASIDE), tc.getHalf
  (Conduit.ASIDE));
```

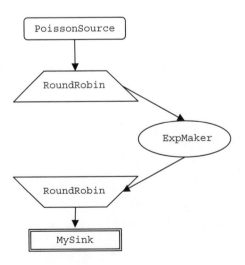

Figure 8.1   The initial architecture of our simulation

Note that, in this code, we ask `PoissonSource` to make 100 packets and inject them according to a Poisson process of intensity $\lambda = 2.0$. Now suppose we want to instantiate and install three `ExponentialBoundedProcess` protocols (say, with maximum queue size 2 and exponential delay $\lambda = 10.0$) in between the `RoundRobin` muxes. We can do this by sending three `InitializeVisitors` into the system:

```
for (int i=0; i<3; i++) {
  InitializeVisitor v =
    new InitializeVisitor(i, 10.0, true, 2);
  rc1.acceptVisitorFrom(v, Conduit.ASIDE);
}
```

These `InitializeVisitors` will hit the `RoundRobin` accessor, which will be unable to route them, as per the code for `getNextConduit()` in the Round-Robin class. Thus, the visitors will get routed to the `ExpMaker` factory, where they will result in the creation of new `ExponentialBoundedProcess` protocols (in `ExpMaker`'s `create()` method). This will get properly installed between the two `RoundRobin` muxes, as per the default factory/mux behaviors. The net outcome will be an architecture as shown in Figure 8.2.

Now when `PoissonSource` begins to eject its `PacketVisitors` out to the upper `RoundRobin` mux, the mux will route (i.e., demultiplex) these visitors in a round-robin way to each of the three `ExponentialBoundedProcess` protocols. Each of these protocols will take a random amount of time to "process" the `PacketVisitor` (with the time depending on an exponential distribution).

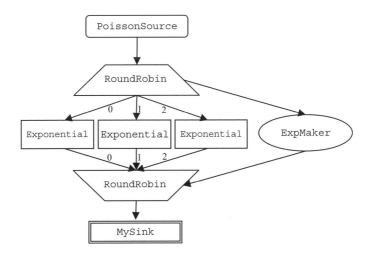

Figure 8.2 Proposed simulation architecture

Having processed the `PacketVisitor`, each `ExponentialBounded-Process` protocol will forward the visitor to the lower `RoundRobin` mux, which will route the visitors (i.e., multiplex) to the `MySink` terminal.

Let us look at the output of a sample execution of this system, which asks `PoissonSource` to make 100 packets and inject them according to a Poisson process of intensity $\lambda = 2.0$:

```
exp-factory making EBP, hooking it up
bexp-0 received visitor Initialize[0] at time 0.0 ... fromA_Q
  has size: 1
exp-factory making EBP, hooking it up
bexp-1 received visitor Initialize[1] at time 0.0 ... fromA_Q
  has size: 1
exp-factory making EBP, hooking it up
bexp-2 received visitor Initialize[2] at time 0.0 ... fromA_Q
  has size: 1

sink received visitor Initialize[1] at time 7.0
bexp-0 received visitor Packet[1]:0:(0) at time 10.0 ...
  fromA_Q has size: 2
sink received visitor Initialize[0] at time 11.0
bexp-1 received visitor Packet[2]:0:(0) at time 12.0 ...
  fromA_Q has size: 1
bexp-2 received visitor Packet[3]:0:(0) at time 13.0 ...
  fromA_Q has size: 2
bexp-0 received visitor Packet[4]:0:(0) at time 14.0 ...
  fromA_Q has size: 2
sink received visitor Initialize[2] at time 15.0
bexp-1 received visitor Packet[5]:0:(0) at time 17.0 ...
  fromA_Q has size: 2
sink received visitor Packet[2]:0:(0) at time 19.0

            <many lines of output>

bexp-0 received visitor Packet[97]:0:(0) at time 215.0 ...
  fromA_Q has size: 2
bexp-1 received visitor Packet[98]:0:(0) at time 219.0 ...
  fromA_Q has size: 1
bexp-2 received visitor Packet[99]:0:(0) at time 219.0 ...
  fromA_Q has size: 2
sink received visitor Packet[91]:0:(0) at time 220.0
```

Table 8.1 Performance metrics from our simulation

Measure/protocol	0	1	2
Max queue size	2.0	2.0	2.0
Mean queue size	1.74	1.69	1.71
Dropped packets	3	4	3

```
sink received visitor Packet[96]:0:(0) at time 221.0
bexp-0 received visitor Packet[100]:0:(0) at time 223.0 ...
   fromA_Q has size: 2
sink received visitor Packet[97]:0:(0) at time 226.0
sink received visitor Packet[99]:0:(0) at time 229.0
sink received visitor Packet[100]:0:(0) at time 229.0
sink received visitor Packet[98]:0:(0) at time 230.0

SUMMARY bexp-0: fromA, total 32, max 2.0, mean=
1.7478260869565216, dropped 3
SUMMARY bexp-0: fromB, total 0, max 0.0, mean=0.0, dropped 0
SUMMARY poisson source: sent 100 packets, received 0 packets
SUMMARY roundrobin-top: recvd 100 packets
SUMMARY exp-factory: made 3 Exps
SUMMARY bexp-2: fromA, total 30, max 2.0, mean=
1.6956521739130435, dropped 4
SUMMARY bexp-2: fromB, total 0, max 0.0, mean=0.0, dropped 0
SUMMARY bexp-1: fromA, total 31, max 2.0, mean=
1.7130434782608697, dropped 3
SUMMARY bexp-1: fromB, total 0, max 0.0, mean=0.0, dropped 0
SUMMARY roundrobin-bottom: recvd 0 packets
SUMMARY sink: received 93 packets
```

We can see the maximum and average queue sizes in each of the three Exponential BoundedQueues, as well as the number of packets that were dropped by each one as shown in Table 8.1.

8.7 Summary

In this chapter, we created some useful network simulation elements that incorporate temporal constructs (using the FDES) with dataflow constructs (using CASiNO). These blocks, included a Poisson packet source and exponential delay with bounded buffer capacity. These components will serve as elements within our exposition of queuing theory in Chapter 9.

9

Queuing Theory

Queuing theory is important in simulations because it provides theoretical predictions against which to validate simulation results. In other words, often a simulation model may be simplified so that queuing theory may be applied to yield theoretical predictions. These predictions are then compared to the experimental measurements obtained through computation to give some assurance of the validity of the simulation implementation.

This chapter presents a brief discussion on several topics of queuing theory. In the first part, basic concepts will be covered. In the second part, specific practical cases will be presented and discussed. Whenever possible, there will be code samples implemented using the CASiNO framework (developed in Chapters 6 and 8), the SimJava Package, or MATLAB. It is expected that the reader is familiar with SimJava and MATLAB in order to follow the examples which will appear later in this chapter.

In the first part of the chapter, the following topics will be covered:

- Introduction to stochastic processes
- Discrete-time and continuous-time Markov chains
- Some properties of Markov chains
- Chapman–Kolmogorov equation (C–K equation)
- Birth–death process (BD process).

Then, in the second part, we will cover:

- Little's theorem
- The exponential distribution
- The Poisson process
- Standard notation of queuing systems

Network Modeling and Simulation   M. Guizani, A. Rayes, B. Khan and A. Al-Fuqaha
© 2010 John Wiley & Sons, Ltd.

- $M/M/1$ queue
- $M/M/n$ queue
- $M/M/1/b$ queue
- $M/M/m/m$ queue.

9.1 Introduction to Stochastic Processes

A *stochastic process* (or a *random process*) is a family of random variables indexed by an ordered set T, formally expressible as $\{X(t), t \in T\}$. That is, for any fixed t in T, $X(t)$ is a random variable. A stochastic process is the counterpart to a deterministic system in probability theory. Instead of dealing with only one possible event determining how the system might evolve over time, in stochastic processes there is some indeterminacy in its future evolution (the values of $X(t)$ for higher values of t in T)—though this indeterminacy is quantifiable with probability distributions. Stochastic processes are classified in different ways, such as by index set, and by the state space, or the "set of all possible values that the random variables $X(t)$ can assume." The following are a few examples of the different cases that can occur.

Example 1   Let W be the amount of time that elapses from the arrival of a message till the processing begins, $\{W(t), t \geq 0\}$. The random variable is indexed by the arrival time of the message (t), which is a continuous parameter, and the state space of $W(t)$ is also continuous since the waiting time can be any real non-negative number.

Example 2   Let $X_n, n = \{3.2, 2.1, 5.0, 4.3, 7.77, 1.2, 6.44\}$, denote the average time to run a job at a computer center on the nth day of the week. That is, X_1 is the average job time on day 1: Sunday; X_2 is the average job time on day 2: Monday; and so on. Then $\{X_n\}$ is a discrete set, though the state space is continuous.

Example 3   If the number of packets (P) in a queue of messages at the time when a message arrives is $\{P(t), t \geq 0\}$, then the random variable is indexed by the arrival time of the message (t). This is considered to be a continuous parameter and the state space is discrete, since the number of messages in the queue is always an integer.

Example 4   Let $X_n, n = \{0, 2, 5, 4, 7, 1, 0\}$, denote the number of cars that passed through checkpoint A on the nth day of the week. That is, X_1 is the number on day 1: Sunday; X_2 is the number on day 2: Monday; and so on. Then $\{X_n\}$ is a discrete set, and the state space is discrete as well. Figure 9.1 shows the different cases that could arise as random variables.

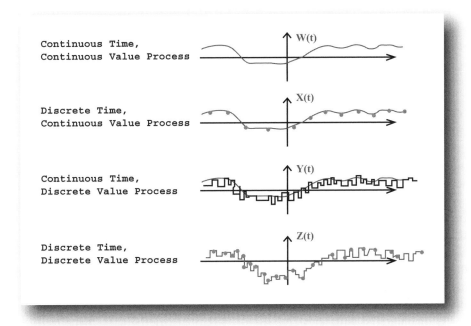

Figure 9.1   Different cases of random variables

A random process can be used to generate graphs of natural events, like the amount of rainfall as a function of the index variable, time. A hypothetical graph showing how much rain fell in the city of Chicago during 2006 with index variable $1 \leq t \leq 365$ days is shown in Figure 9.2.

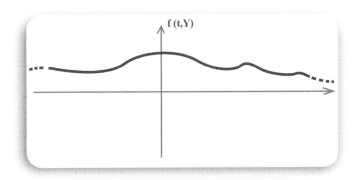

Figure 9.2   Rainfall in Chicago during 2006 ($1 \leq t \leq 365$ days)

If there are only countably many times (i.e., values of the index variable) at which the random variable can change its value, the random process is called a *discrete parameter process* X_n. If these changes occur anywhere within a continuous interval, we have a *continuous parameter process* $X(t)$; we often write X_n to refer to a *random or stochastic sequence* and $X(t)$ to a *stochastic or random process*.

The most common types of stochastic processes that appear in the context of simulation studies are listed below. They are characterized by different kinds of dependency relations among their constituent random variables. This classification provides a global overview of the different sub-branches within the field of queuing theory.

- **Independent processes:** These are the simplest and most trivial. They consider the sequence in which $\{X_n\}$ forms a set of independent random variables.
- **Markov processes:** A set of random variables, say $\{X_n\}$, forms a Markov chain if the probability that the next state is X_{n+1} depends only upon the current state X_n and not upon any earlier values.
- **Birth–death processes:** A Markov process with discrete state space, whose states can be enumerated with an index, say $i = 0, 1, 2 \ldots$ such that the state transitions occur *only* between neighboring states.
- **Semi-Markov processes:** These extend the notion of Markov processes by allowing the time between state transitions to obey an arbitrary probability distribution. The case when a transition must be made at every unit time is the classical Markov process, in which the time spent in one state is *geometrically distributed.*
- **Random walks:** Can be thought of as an object moving among states in some state space (e.g., discrete space). A *location* of this object is defined in a certain state.
- **Stationary processes:** A stochastic process $X(t)$ is said to be stationary if there is a $F_x(x; t)$ which is *invariant* to the shifts in time for all values of its arguments.

A Markov process with a discrete state space is referred to as a *Markov chain*. A set of random variables, say $\{X_n\}$, forms a Markov chain if the probability that the next state is X_{n+1} depends *only* upon the current state X_n and not upon any previous values. This also implies that the probability distribution of X_n (denoted P_n) depends only on the previous state and no earlier X_i (for $i < n - 1$). In short, we have a random sequence in which the dependency is extended backward at most one unit in time.

Consider a stochastic process $\{X(t), t \in T\}$. This stochastic process is called a Markov process if for any increasing sequence of $n + 1$ values of (say time) t such that $t_1 < t_2 < t_3 < \cdots < t_n < t_{n+1}$ in the index set, and any set $\{x_1, x_2, x_3, \ldots, x_{n+1}\}$ of

Table 9.1 Markov process classifications

Types of parameters	State space	
	Discrete	Continuous
Discrete time	Discrete-time Markov chain	Discrete-time Markov process
Continuous time	Continuous-time Markov chain	Continuous-time Markov process

$n + 1$ states, we have:

$$P[X(t_{n+1}) = x_{n+1}|X(t_1) = x_1, X(t_2) = x_2, X(t_3) = x_3, \ldots, X(t_n) = x_n]$$
$$= P[X(t_{n+1}) = x_{n+1}|X(t_n) = x_n].$$

This indicates that the future of the process depends only upon the present state and *not* upon the history of the process. So, we can say that the entire history of the process is summarized in the present state. Markov processes can be further classified as shown in Table 9.1.

9.2 Discrete-Time Markov Chains

In discrete-time Markov chains, a variable can take one of a denumerable set of values and is permitted to transition between these values only at discrete times, i.e., at t_n, $n = 0, 1, 2, 3, 4 \ldots$. A discrete-time Markov chain might be presently in some state, say i, so when $t = t_n$, $X_n = i$. It then makes a state transition at $t = t_{n+1}$ to $X_{n+1} = j$. In general, the probabilities of one-step transitions at step $n + 1$ can be defined as follows:

$$P[X_{n+1} = j|X_n = i] \quad \text{where } n = 0, 1, 2, 3 \ldots.$$

Because these probabilities are independent of n, each probability can actually be denoted P_{ij}. In this case, the Markov chain is said to have *stationary transition probabilities* or to be *homogeneous in time*. The transition probabilities can be represented as a square table, which is called the *transition probability matrix* of the chain. If the number of states is finite (n), then the matrix is $n \times n$ in size, otherwise the matrix is infinite. The transition probability matrix **P** is shown below. Note that in the matrix, $P_{ij} \geq 0$, $i, j = 0, 1, 2 \ldots$, and $\sum_{j=0}^{\infty} P_i = 1$, $i = 0, 1, 2$; an example of such a matrix is shown in Figure 9.3.

Example 5   In Bernoulli trials, the probability of success on each trial is p $(0 < p < 1)$ and the probability of failure is $q = 1 - p$. If we represent the longest consecutive

$$
P = \begin{bmatrix}
P_{00} & P_{01} & P_{02} & P_{03} & \cdots \\
P_{10} & P_{11} & P_{12} & P_{13} & \cdots \\
P_{20} & P_{21} & P_{22} & P_{23} & \cdots \\
\vdots & \vdots & \vdots & \vdots & \\
P_{10} & P_{11} & P_{12} & P_{13} & \cdots
\end{bmatrix}
$$

Figure 9.3 Transition probability matrix

sequence of successes as a random variable X, and if the first six outcomes observed from a number of Bernoulli trials are success, failure, success, success, success, then failure (and we use 1 to represent success and 0 to represent failure), then we would have $X_0 = 1$, $X_1 = 0$, $X_2 = 1$, $X_3 = 2$, $X_4 = 3$, and $X_5 = 0$. The transition matrix is shown in Figure 9.4.

$$
P_{ij} = \begin{bmatrix}
q & p & 0 & 0 & 0 & 0 & \cdots \\
q & 0 & p & 0 & 0 & 0 & \cdots \\
q & 0 & 0 & p & 0 & 0 & \cdots \\
q & 0 & 0 & 0 & p & 0 & \cdots \\
q & p & 0 & 0 & 0 & p & \cdots \\
\vdots & \vdots & \vdots & \vdots & \vdots & \vdots &
\end{bmatrix}
$$

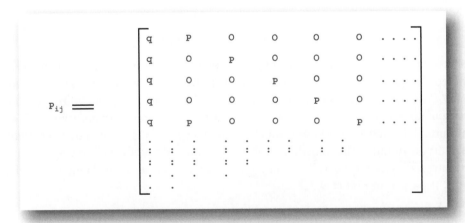

Figure 9.4 Transition matrix for Example 9.5

9.3 Continuous-Time Markov Chains

In a continuous-time Markov chain, the index set T is continuous, and hence the random variable is able to change at any point in time, so a stochastic process $\{X(t), t \in T\}$ is formed. This stochastic process is called a Markov process if for any increasing sequence of $n + 1$ values of (say, time) t such that $t_1 < t_2 < t_3 \ldots < t_n < t_{n+1}$ in the index set, and any set $\{x_1, x_2, x_3, \ldots, x_{n+1}\}$ of $n + 1$ states, we have:

$$P[X(t_{n+1}) = x_{n+1}|X(t_1) = x_1, X(t_2) = x_2, X(t_3) = x_3, \ldots, X(t_n) = x_n]$$
$$= P[X(t_{n+1}) = x_{n+1}|X(t_n) = x_n].$$

9.4 Basic Properties of Markov Chains

Reducibility: A state j is said to be *reachable* from a different state i (written $i \rightarrow j$) if, given that we are still in state i, there is a probability ≥ 0 that at some time in the future the system will be in state j. Also one can say that state j is *accessible* from state i if there exists an integer $n \geq 0$ such that $P(X_n = j|X_0 = i) > 0$. In this case, we say that state i is able to *communicate* with state j. The transitive closure of the relation "is able to communicate with" defines an equivalence relation on the set of states, and a Markov chain is said to be *irreducible* if this equivalence relation consists of exactly one class.

Periodicity: A state i is defined to have period M if any return to state i occurs in multiples of M time steps. So the period of a state is defined as:

$$M = \text{Greater\_Common\_Divisor } \{n : P(X_n = I|X_0 = i) > 0\}.$$

If $M = 1$, the state is said to be aperiodic.

Ergodicity: A state i is said to be *transient* if, given that we start in state i, there is a non-zero probability that we will never return to i. If a state i is not transient, then if the expected return time is finite, the state is called *positive recurrent*. A Markov chain is called *ergodic* if all its states are aperiodic and positive recurrent.

Regularity: A Markov chain is said to be regular if there is a number n such that:

$$(P_{ij})^{(n)} > 0$$

for all i, j. Otherwise, the chain is irregular.

Memoryless: A discrete Markov chain is said to be *memoryless* if its state times are geometrically distributed. A continuous-time Markov chain is said to be *memoryless* if its state times are exponentially distributed.

9.5 Chapman–Kolmogorov Equation

The Chapman–Kolmogorov (C–K) equation can be used to compute the n-step transition probabilities of a Markov chain. Since a Markov chain is homogeneous in time, if the one-step transition probabilities on it are applied (i.e., $P(A_n)$ given A_{n-1}), the C–K equation provides a relation for multiple step transitions:

$$P_{ij}^{n+m} = \sum_{k=0}^{\infty} P_{ik}^{n} P_{kj}^{m} \quad \text{for all } i, j, m, n \geq 0.$$

In particular:

$$P_{ij}^{n} = \sum_{k=0}^{\infty} P_{ik}^{n-1} P_{kj} \quad \text{when } n = 2, 3, 4 \ldots.$$

Example 6   Assume that a staged communication device receives a sequence of bibinary digits (0 or 1). Inside the device, each received digit is passed sequentially through a sequence of stages. At the end of each stage, the probability that the digit will be transmitted (uncorrupted) to the next stage of the device is 70%. What is the probability that a digit "0" is received as "0" at the fifth stage? Here, we have to find P_{00}^{4}, because we count the n stages from 0 to 4, so the state transition matrix **P** is the matrix shown in Figure 9.5.

Note that the C–K equation has different forms depending on what type of Markov chain is being considered. Although in most applications of simulation, the homogeneous discrete-time Markov chain is assumed, for completeness Table 9.2 shows a summary of the C–K equation forms.

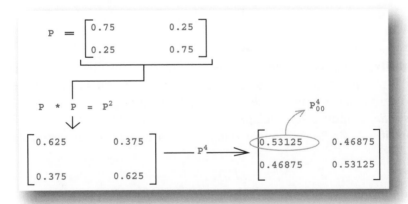

Figure 9.5   State transition matrix for Example 6

Table 9.2 Summary of C–K equation

	Discrete time		Continuous time	
	Homogeneous	Non-homogeneous	Homogeneous	Non-homogeneous
C–K equation	$P_{ij}^n = \sum_{k=0}^{\infty} P_{ik}^{n-m} P_{kj}^m$	$P_{ij}^{n+m} = \sum_{k=0}^{\infty} P_{ik}^{n+q} P_{kj}^{q+m}$	$P_{ij}^t = \sum_{k=0}^{\infty} P_{ik}^{t-s} P_{kj}^s$	$P_{ij}^{s+u} = \sum_{k=0}^{\infty} P_{ik}^{s+u} P_{kj}^{u+t}$

9.6 Birth–Death Process

A birth–death (BD) process refers to a Markov process with discrete state space, the states of which can be enumerated with index, say, $i = 0, 1, 2\ldots$ such that the state transitions can occur *only* between neighboring states, i.e., either $i \rightarrow i + 1$ or $i \rightarrow i - 1$. The concept of a BD process arises most frequently in the context of *population*, e. g., the number of customers in the queuing system. In this, the notion of *death* is a customer's departure from the system (service done), and the notion of *birth* is a customer's arrival at the system (within an arrival rate). The size of the population will change by (at most) 1 at each time step; size can be increased by one "birth," or decreased by one "death." Note that when the system is in a certain state, it can be denoted by E_k, such that E is the state which consists of k members. The transition probabilities P_{ij} do not change with time, and can be described diagrammatically as shown in Figure 9.6,

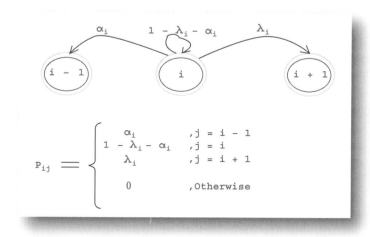

Figure 9.6   State transition diagram and resulting probability of the BD process

where:

- $\lambda_i =$ birth (increase population by 1);
- $\alpha_i =$ death (decrease population by 1).

Several special cases exist:

- $\lambda_i > 0$ (birth is allowed);
- $\alpha_0 = 0$ (no death);
- pure birth (means no decrement, only increment);
- pure death (means no increment, only decrement).

Example 7 (a BD process and memoryless Markov chains):   If a BD process has a uniform transition probability p from state i to state $i + 1$ (independent of i), and a uniform probability for the reverse transition, then

- P[system in state i for N time units | system in current state i] $= P^N$
- P[system in state i for N time units before exiting from state i] $= P^{N*} (1 - p)$.

It can be shown that the state times are geometrically distributed. Thus, such a BD process is memoryless.

In the second part of this chapter, some of the key properties of the exponential distribution and the Poisson process that are commonly used in the simulation of data propagation across network links are presented. However, Little's theorem is introduced first. Then the standard (Kendall's) notation of queuing systems is discussed and a description of the most prominent and basic queuing systems are given. These are:

- $M/M/1$ queue
- $M/M/n$ queue
- $M/M/1/b$ queue
- $M/M/m/m$ queue.

9.7 Little's Theorem

Little's theorem (or Little's law) is a very useful law in studying queuing systems. It can be shown that this law holds for almost every steady-state queuing system. It can be shown that when there is a crowded system (i.e., a large number of customers, $N \gg 0$) that is associated with long customer delays ($T \gg 0$), Little's theorem can be written as $N = \lambda T$. This means that the number of customers in the system is equal to the arrival rate (λ) times the total time T a customer spends in the system to get serviced, where N is the random variable for the number of jobs or customers in the system, λ is the arrival rate at which jobs arrive, and T is the random variable for the time a job spends in the system.

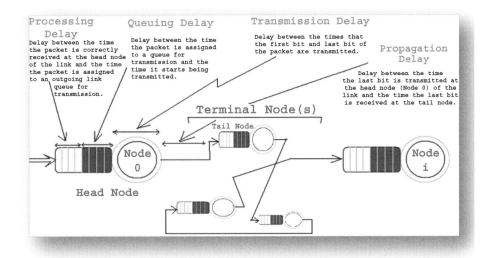

Figure 9.7 Possible packet delays

9.8 Delay on a Link

The packet delay is the sum of delays on each link (or subnet link) that was traversed by the packet. Link delay consists of four delays that face a packet arriving at a queue for processing. These are: time to process the packet, time spent in the queue, transmission time, and propagation time. These are shown and defined in Figure 9.7.

When packets flow on a link, the exponential distribution is used to model inter-arrival times when arrivals are completely random. It is also used to model packet processing delay time. The Poisson process is important because it is used to model the arrival of customers, packets, etc. The key parameter that one should know when dealing with a system which contains a Poisson process is λ, the average arrival rate of customers. The number of arrivals in any interval, say $[t_0, t_0 + T]$, is Poisson distributed with intensity λT, where λ is the average number of arrivals in a unit interval of time. Note that the mean and variance of inter-arrival times are $1/\lambda$ and $1/\lambda^2$, respectively. However, the inter-arrival times are independent and exponentially distributed with parameter λ.

9.9 Standard Queuing Notation

We now introduce a very useful notation to represent queue models based on their structure. This is known as Kendall's notation. The notation consists of four parameters separated by slashes (/). The meaning of parameters and what possible values these can take are summarized in Figure 9.8.

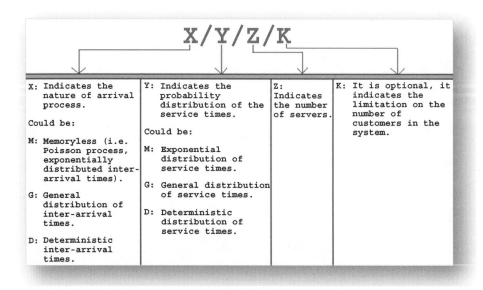

Figure 9.8   Standard queuing notation (Kendall's notation)

Next, some of the more common configurations of queues are considered. These are:

- *M/M/*1 queue
- *M/M/n* queue
- *M/M/*1/*b* queue
- *M/M/m/m* queue.

Each of these will be implemented and analyzed in one of several ways, using a model of the CASiNO simulation framework implemented, designed, and discussed in Chapters 6 and 8, SimJava, or MATLAB. So, it is highly recommended that the reader should be familiar with both queuing theory as well as the various implementation platforms used here.

9.10 The *M/M/*1 Queue

This model assumes a random Poisson arrival process and a random exponential service time distribution. The arrival rate does not depend on the number of customers already in the system (so customers are not discouraged from arriving because the queue is full). Imagine the state of the *M/M/*1 queue being implemented based on a BD process whose transitions are shown in Figure 9.9.

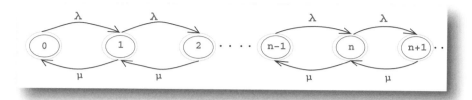

Figure 9.9 State transition diagram of the *M/M/*1 system

Analytically, the *M/M/*1 queue exhibits the following properties:

- The utilization factor (which is the proportion of time the server is being busy) is $\rho = \lambda/\mu$.
- The average number of customers in the system is $N = \rho/(1 - \rho)$.
- The average waiting time in the queue is $W = T - 1/\mu = \rho/(\mu - \lambda)$.
- The average customer time in the system is $T = N/\lambda = 1/(\mu - \lambda)$.
- The average number of customers in the queue is $N = \lambda W = \rho^2/(1 - \rho)$.

Within more complex real-life simulation models, the *M/M/*1 queue may frequently arise as a substructure. The above formal properties may then be used to validate the simulation results, and (in the case of steady-state inputs) can serve as justification for further simplifications of the model.

9.10.1 A CASiNO Implementation of the M/M/1 Queue

Based on the classes designed in the case study of CASiNO in Chapter 8, the CASiNO code implementation below is of an *M/M/*1 queuing system. It consists of a `Poisson Source` which acts as a source of packets, an `ExponentialProcess` (exponential service time) packet processor, and a `MySink` module, all connected in a row as shown in the flow chart of Figure 9.10.

The code that creates this is as follows:

```
public class Run_MM1 {

  public static void main (String [ ] args) {

    if (args.length != 3) {
      System.out.println("Arguments: numpackets "+
                         "source-intensity processing-
                         exp");
```

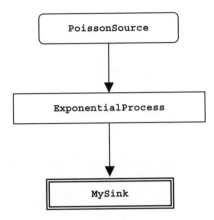

Figure 9.10 Flow chart of the CASiNO simulation architecture

```
    System.exit(-1);
  }
int numpackets = Integer.parseInt(args[0]);
double source = Double.parseDouble(args[1]);
double processing = Double.parseDouble(args[2]);

  Conduit sc = new Conduit("poisson source",
                        new PoissonSource(numpackets,
                        source));
  Conduit mc = new Conduit("exp", new ExponentialProcess
    (processing));
  Conduit tc = new Conduit("sink", new MySink());

Conduit.join(sc.getHalf(Conduit.ASIDE), mc.getHalf
  (Conduit.ASIDE));
Conduit.join(mc.getHalf(Conduit.BSIDE), tc.getHalf
  (Conduit.ASIDE));

  Thread t = new Thread(Scheduler.instance());
  t.start();
  try { t.join(); }
  catch (Exception e) { }

  CASINO.shutdown();
  }
}
```

The program takes three arguments:

- the number of packets;
- the intensity of the Poisson process that governs the generation of packets in the source;
- the intensity of the exponential distribution that governs the packet processor's time to process a packet.

If we run this program once, specifying that 1000 packets are injected at a Poisson intensity of 2.0, with the intensity of the exponential processing delay being 10.0, we get the following results:

```
SUMMARY sink: received 1000 packets
SUMMARY poisson source: sent 1000 packets, received 0 packets
SUMMARY exp: fromA, total 1000, max 11.0, mean=
  5.411219512195122
SUMMARY exp: fromB, total 0, max 0.0, mean=0.0
```

So we observe that the maximum queue size for this simulation run was 11, while the mean was 5.41 packets. These numbers can then be validated against the previously described theoretical predictions of the behavior of an *M/M/*1 system.

9.10.2 A SimJava Implementation of the M/M/1 Queue

Let us now use SimJava to create an *M/M/*1 queue model with the required entities: the source of events "customers," the queue, and the server. The classes are shown in Figure 9.11 and summarized all together with a caption of what the animation applet gives as a representation of entities; we end up with one source, one "infinite" queue, and one server. The first class is the Source class, with the arriving events based on a Poisson process, the second class is the Queue class, and the last class is the Server, which provides an exponentially distributed service.

9.10.3 A MATLAB Implementation of the M/M/1 Queue

We now use the package "sim_event" in MATLAB to design an *M/M/*1 queue model, using the required entities with the suitable attributes. Note the type of distribution for the source and how to set the number of servers as shown in Figure 9.12.

```
class Source extends Sim_entity {        (A)
  private Sim_port out;
  private Sim_poisson_obj delay;

  Source(String name, int x, int y,
String image, double mean){
    super(name, image, x, y);
    out = new Sim_port("Out", "port",
            Anim_port.RIGHT, 40);
  add_port(out);
  lambda = new Sim_negexp_obj("Delay", mean);
  add_generator(lambda);
  }
```

```
class Processor extends Sim_entity {    (B)
  private Sim_port in;
  private Sim_negexp_obj delay;
  private Sim_stat stat;

  Processor(String name, int x, int y,
            String image, double mean) {
    super(name, image, x, y);
    in = new Sim_port("In", "port",
        Anim_port.LEFT, 40);
    add_port(in);
    delay = new Sim_negexp_obj("Delay",mean);
    add_generator(delay); // This method is
                          //used to allow
                          //Sim_system to re-
                          //seed the generator.
                          //This is performed
                          //when independent rep-
                          //lications have been
                          //selected as an inout
                          //analysis.

    stat = new Sim_stat();
    stat.add_measure(Sim_stat.UTILISATION);
    set_stat(stat);
    add_param(new Anim_param("Busy",
    Anim_param.STATE, "idle", 0, 0)); //this
    //was added to trace the status of the
    //processor using animation "busy/idle"
  }

  public void body() {
    while (Sim_system.running()) {
      Sim_event e = new Sim_event();
      sim_get_next(e);
      sim_trace(1, "P busy");
      sim_process(delay.sample());
      sim_trace(1, "P idle");
      sim_completed(e);} }}
```

```
class Queue extends Sim_entity {        (C)
  private Sim_port in, out1, out2;
  private Sim_negexp_obj delay;
  private Sim_random_obj prob;
  private Sim_stat stat;

  Queue(String name, int x, int y, String
    super(name, image, x, y);
    in = new Sim_port("In", "port",
        Anim_port.LEFT, 40);
    out1 = new Sim_port("Out1", "port",
        Anim_port.RIGHT, 20);
    add_port(in);
    add_port(out1);
    prob = new Sim_random_obj("Probability");
    stat = new Sim_stat();
    stat.add_measure(Sim_stat.WAITING_TIME)

    set_stat(stat);
  }

  public void body() {
    while (Sim_system.running()) {
      Sim_event e = new Sim_event();
      sim_get_next(e);
      sim_process(Sim_stat.THROUGHPUT);
      sim_completed(e);

      //here we just want to to schedule
      //events to one processor (M/M/1)

        sim_schedule(out1, 0.0, 1);
        sim_trace(1, "S Out1");
    }
  }
}
```

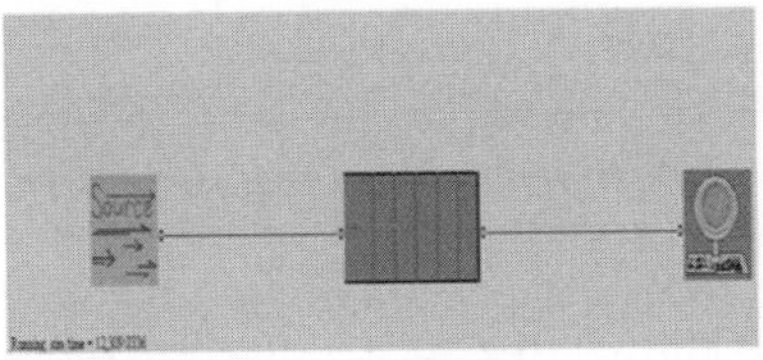

Figure 9.11   SimJava implementation of *M/M/*1 queue

9.11 The *M/M/m* Queue

This model consists of a single queue with *m* servers, and the customers arrive according to a Poisson process with rate λ. Each of the *m* servers has a distribution of service time that is exponential with mean $1/\mu$.

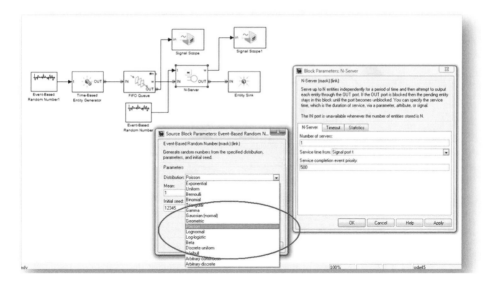

Figure 9.12   A snapshot of MATLAB *M/M/*1 queue implementation

Analytically, the *M/M/m* queue exhibits the following properties:

- The server utilization is measured by $\rho = \lambda/m\mu$, where *m* is the number of servers.
- The average waiting time in a queue of customers is:

$$W = \frac{N_Q}{\lambda} = \frac{\rho p_Q}{\lambda(1-\rho)}.$$

such that the average number of customers in the queue N_Q is:

$$N_Q = \sum_{n=0}^{\infty} n p_{m+n} = \frac{\rho p_Q}{1-\rho}.$$

- The average customer time in the system is:

$$T = \frac{1}{\mu} + W = \frac{1}{\mu} + \frac{p_Q}{m\mu - \lambda}.$$

- The average number of customers in the system is:

$$N = \lambda T = m\rho + \frac{\rho p_Q}{1-\rho}.$$

Within a more complex real-life simulation model, the *M/M/m* queue may frequently arise as a substructure. The above formal properties may then be used to

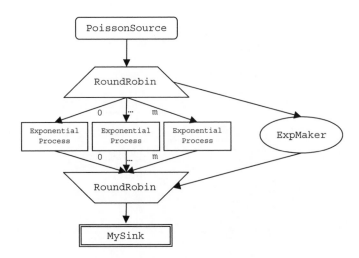

Figure 9.13   Flow chart of the CASiNO implementation of the *M/M/m* queue

validate the simulation results, and (in the case of steady-state inputs) can serve as justification for further simplifications of the model.

9.11.1 A CASiNO Implementation of the M/M/m Queue

Listed below is the CASiNO code implementing an *M/M/m* queuing system using the CASiNO simulation discussed in Chapter 8. It consists of a `PoissonSource` which acts as source of packets, and two `RoundRobin` muxes around a factory in which `ExpMaker` creates `ExponentialProcess` objects dynamically, in response to the arrival of `InitializeVisitors`. The lower `RoundRobin` mux is connected to a `MySink` module where all visitors die. The architecture of the CASiNO simulation is shown in Figure 9.13.

The code that creates this is as follows:

```
public class Run_MMn {

  public static void main (String [ ] args) {
    if (args.length != 4) {
      System.out.println("Arguments: numpackets "+
                         "source-intensity processing-exp
                         numservers");
      System.exit(-1);
    }
    int numpackets = Integer.parseInt(args[0]);
    double source = Double.parseDouble(args[1]);
```

```
double processing = Double.parseDouble(args[2]);
double numservers = Double.parseDouble(args[3]);

   Conduit sc = new Conduit("poisson source",
                            new PoissonSource
                               (numpackets, source));
   Conduit rc1 = new Conduit("roundrobin-top", new Round-
      Robin());
   Conduit fc = new Conduit("exp-factory", new
      ExpMaker());
   Conduit rc2 = new Conduit("roundrobin-bottom", new
      RoundRobin());
   Conduit tc = new Conduit("sink", new MySink());

Conduit.join(sc.getHalf(Conduit.ASIDE), rc1.getHalf
   (Conduit.ASIDE));
Conduit.join(rc1.getHalf(Conduit.BSIDE), fc.getHalf
   (Conduit.ASIDE));
Conduit.join(fc.getHalf(Conduit.BSIDE), rc2.getHalf
   (Conduit.BSIDE));
Conduit.join(rc2.getHalf(Conduit.ASIDE), tc.getHalf
   (Conduit.ASIDE));
for (int i=0; i<numservers; i++) {
   InitializeVisitor v =
      new InitializeVisitor(i, processing, false, 0);
   rc1.acceptVisitorFrom(v, Conduit.ASIDE);
}

Thread t = new Thread(Scheduler.instance());
t.start();
try { t.join(); }
catch (Exception e) { }

CASINO.shutdown();
   }
}
```

The program takes four arguments:

- the number of packets;
- the intensity of the Poisson process that governs the generation of packets in the source;

- the intensity of the exponential distribution that governs the packet processor's time to process a packet.
- number of servers.

If we run this program once, specifying that 1000 packets are injected at a Poisson intensity of 2.0, with the intensity of the exponential processing delay being 10.0, we get the following results.

When the number of servers is 2:

```
SUMMARY exp-0: fromA, total 501, max 7.0, mean=
    3.0590295147573787
SUMMARY exp-1: fromA, total 501, max 7.0, mean=
    2.978489244622311
SUMMARY poisson source: sent 1000 packets, received 0 packets
SUMMARY roundrobin-top: recvd 1000 packets
SUMMARY exp-factory: made 2 Exps
SUMMARY sink: received 1002 packets
```

When the number of servers is 4:

```
SUMMARY exp-0: fromA, total 251, max 3.0, mean=
    1.720347155255545
SUMMARY exp-1: fromA, total 251, max 3.0, mean=
    1.7135969141755063
SUMMARY exp-2: fromA, total 251, max 3.0, mean=
    1.800867888138862
SUMMARY exp-3: fromA, total 251, max 3.0, mean=
    1.6938283510125363
SUMMARY poisson source: sent 1000 packets, received 0 packets
SUMMARY roundrobin-top: recvd 1000 packets
SUMMARY exp-factory: made 4 Exps
SUMMARY sink: received 1004 packets
```

When the number of servers is 8:

```
SUMMARY poisson source: sent 1000 packets, received 0 packets
SUMMARY exp-0: fromA, total 126, max 2.0, mean=1.1220703125
SUMMARY exp-1: fromA, total 126, max 2.0, mean=1.09130859375
SUMMARY exp-2: fromA, total 126, max 2.0, mean=1.1494140625
SUMMARY exp-3: fromA, total 126, max 2.0, mean=1.08349609375
SUMMARY exp-4: fromA, total 126, max 2.0, mean=1.072265625
```

```
SUMMARY exp-5: fromA, total 126, max 2.0, mean=1.0986328125
SUMMARY exp-6: fromA, total 126, max 2.0, mean=1.07080078125
SUMMARY exp-7: fromA, total 126, max 2.0, mean=1.10205078125
SUMMARY roundrobin-top: recvd 1000 packets
SUMMARY exp-factory: made 8 Exps
SUMMARY roundrobin-bottom: recvd 0 packets
SUMMARY sink: received 1008 packets
```

These numbers can then be validated against the previously described theoretical predictions of the behavior of an *M/M/n* system.

9.11.2 A SimJava Implementation of the M/M/m Queue

Let us now use SimJava to create an *M/M/m* queue model with the required entities: the code for the `Queue` class and a main class named `ProcessorSubsystem` are all provided below. The capture of the animation applet output as a representation of entities is shown in Figure 9.14.

The simulation has one source, one "infinite" queue, and three servers (*m* = 3). The `Source` and `Server` classes will remain the same as the *M/M/*1 implementation, but now the `Queue` class has multiple outputs. In the simulation running method, we

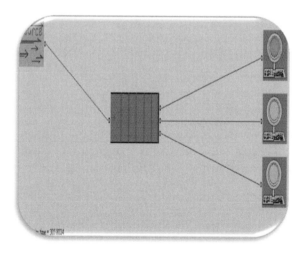

Figure 9.14   A snapshot of the SimJava implementation of the *M/M/m* queue

create three servers and assign the required attributes to them:

```
public class ProcessorSubsystem extends Anim_applet {

  // Set up the animation
  public void anim_layout() {
    //The source's delay between jobs will be exponentially
      distributed
    Source source = new Source("Source", 0, 0, "source", 1);
    Queue queue = new Queue("buffer", 300, 120, "cpu");
      //The Server's distribution will have different means
        and variances
    Server server1 = new Server("server1", 800, 20, "disk1",
      0.3);
    Server server2 = new Server ("server2", 800, 120, "disk1",
      0.3);
    Server Server3 = new Server ("server3", 800, 220, "disk1",
      0.1);

    //linking...
    Sim_system.link_ports("Source", "Out", " buffer ",
      "In");
    Sim_system.link_ports("buffer", "Out1", "server1",
      "In");
    Sim_system.link_ports("buffer ", "Out2", "server2",
      "In");
    Sim_system.link_ports("buffer ", "Out3", "server3",
      "In");
  }
}
```

The running method for the simulation is the same for all the models we are using here. The code here is used to show the results of the simulation with comments on each type of detail to be shown after the simulation is complete:

```
public void sim_setup() {
  //finish after 10000 simulation time units
  Sim_system.set_transient_condition(Sim_system.TIME_
    ELAPSED, 1000);
  Sim_system.set_termination_condition(Sim_system.
    EVENTS_COMPLETED, "Processor", 0, 100, false);
```

```
  // produce 95% confidence intervals
  Sim_system.set_report_detail(true, true);
  Sim_system.generate_graphs(true);
}
  // Choose animation output detail
  public void anim_output() {
    generate_report(true);
    generate_messages(true);
}
```

The new multiple-output Queue class is:

```
class Queue extends Sim_entity {
  ...
  Queue (String name, int x, int y, String image)
    super(name, image, x, y);
    in = new Sim_port("In", "port", Anim_port.LEFT, 40);
    out1 = new Sim_port("Out1", "port", Anim_port.RIGHT, 20);
    out2 = new Sim_port("Out2", "port", Anim_port.RIGHT, 40);
    out3 = new Sim_port("Out3", "port", Anim_port.RIGHT, 60);

    add_port(in);
    add_port(out1); // for server 1
    add_port(out2); // for server 2
    add_port(out3); // for server 3

    prob = new Sim_random_obj("Probability");
    ...
  }

  public void body() {
    while (Sim_system.running()) {
      ...
      sim_trace(1, "S Out1"); //schedule to server 1
        sim_get_next(e);
        sim_process(Sim_stat.THROUGHPUT);
        sim_completed(e);

        //schedule to server 2
        sim_schedule(out2, 0.0, 1);
        sim_trace(1, "S Out2");
```

```
// finally, 3rd server
sim_get_next(e);
sim_process(Sim_stat.THROUGHPUT);
sim_completed(e);

sim_schedule(out3, 0.0, 1);
sim_trace(1, "S Out3"); ...
    }
}
```

9.11.3 A MATLAB Implementation of the M/M/m Queue

Here we use the "sim_event" package of MATLAB to design an *M/M/m* queue model, setting the required attributes of each entity. Figure 9.15 shows how we choose the type of distribution for the source, and how to set the number of servers to 3.

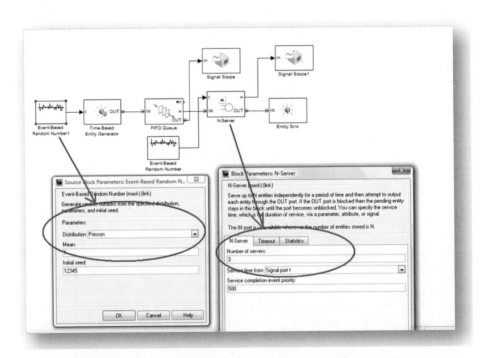

Figure 9.15   A snapshot of MATLAB *M/M/m* queue implementation

9.12 The *M/M/1/b* Queue

This queue is a more realistic extension of the *M/M/1* queue. It assumes that there is a limit to the size of queue "or buffer" which makes it possible to refuse receiving any extra customers when the limit b is exceeded. Arrivals continue to occur according to a Poisson process, and customer service time is exponentially distributed. There is just one server used with a finite queue length of size b.

Analytically, the *M/M/1/b* queue exhibits the following properties:

- The server throughput is:

$$\frac{1 - \rho^b}{1 - \rho^{b+1}} \lambda.$$

- The average number of customers in the system is:

$$N = \frac{\rho}{1 - \rho} - \frac{(b+1)\rho^{b+1}}{1 - \rho^{b+1}}.$$

- The average time spent by a customer in the system (queue time + processing time) is:

$$T = \frac{\text{average number of customers in system}}{\text{service throughput}}.$$

- The blocking probability (the probability that a customer will face a full queue and gets rejected) is:

$$P_b = \frac{(1 - \rho)\rho^b}{1 - \rho^{b+1}}.$$

- The average queue length (the number of customers in the queue) is:

$$L_Q = \frac{\rho}{1 - \rho^{b+1}} \left[\frac{(1 - \rho^b)}{1 - \rho} - b\rho^b \right].$$

Within a more complex real-life simulation model, the *M/M/1/b* queue may frequently arise as a substructure. The above formal properties may then be used to validate the simulation results, and (in the case of steady-state inputs) can serve as justification for further simplifications of the model.

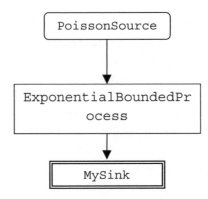

Figure 9.16   Flow chart of the CASiNO implementation of the *M*/*M*/1/*b* queue

9.12.1 A CASiNO Implementation of the M/M/1/b Queue

Listed below is the CASiNO code implementing an *M*/*M*/1/*b* queuing system using the CASiNO simulation discussed in Chapter 8. It consists of a `PoissonSource` which acts as a source of packets, an `ExponentialBoundedProcess` (exponential service time, bounded queue) packet processor, and a `MySink` module, all connected in a row. The architecture of the CASiNO simulation is shown in Figure 9.16.

The code for this is as follows:

```
public class Run_MM1b {

  public static void main (String [ ] args) {

    if (args.length != 4) {
      System.out.println ("Arguments: numpackets source-
        intensity"+" processing-exp bound");
      System.exit(-1);
    }
    int numpackets = Integer.parseInt (args[0]);
    double source = Double.parseDouble (args[1]);
    double processing = Double.parseDouble (args[2]);
    int bound = Integer.parseInt (args[3]);

      Conduit sc = new Conduit ("poisson source",
                   new PoissonSource (numpackets, source));
      Conduit mc = new Conduit ("bexp",
```

```
                   new ExponentialBoundedProcess
                      (processing, bound));
       Conduit tc = new Conduit("sink", new MySink());

     Conduit.join(sc.getHalf(Conduit.ASIDE), mc.getHalf
        (Conduit.ASIDE));
     Conduit.join(mc.getHalf(Conduit.BSIDE), tc.getHalf
        (Conduit.ASIDE));
     Thread t = new Thread(Scheduler.instance());
     t.start();
     try { t.join(); }
     catch (Exception e) { }

     CASINO.shutdown();
   }
}
```

The program takes four arguments:

- the number of packets;
- the intensity of the Poisson process that governs the generation of packets in the source;
- the intensity of the exponential distribution that governs the packet processor's time to process a packet;
- the bound on the packet processor's queue size.

If we run this program once, specifying that 1000 packets are injected at a Poisson intensity of 2.0, with the intensity of the exponential processing delay being 10.0, we get the following results.

When the queue bound is 10:

```
SUMMARY sink: received 994 packets
SUMMARY bexp: fromA, total 994, max 10.0, mean=
  5.442190669371197, dropped 6
SUMMARY bexp: fromB, total 0, max 0.0, mean=0.0, dropped 0
SUMMARY poisson source: sent 1000 packets, received 0 packets
```

When the queue bound is 5:

```
SUMMARY sink: received 759 packets
SUMMARY bexp: fromA, total 759, max 5.0, mean=
  4.354450261780105, dropped 241
```

```
SUMMARY bexp: fromB, total 0, max 0.0, mean=0.0, dropped 0
SUMMARY poisson source: sent 1000 packets, received 0 packets
```

When the queue bound is 2:

```
SUMMARY sink: received 340 packets
SUMMARY bexp: fromA, total 340, max 2.0, mean=
  1.94634402945818, dropped 660
SUMMARY bexp: fromB, total 0, max 0.0, mean=0.0, dropped 0
SUMMARY poisson source: sent 1000 packets, received 0 packets
```

These numbers can then be validated against the previously described theoretical predictions of the behavior of an *M/M/1/b* system.

9.12.2 A SimJava Implementation of the M/M/1/b Queue

Let us use SimJava now to create an *M/M/1/b* queue model with the required entities. The Queue class is the only one provided below because it is the only one that has a major change. Using a while statement, we can keep polling on the situation of the queue until, at some time, "maybe" it gets full. In such a case, no further action, "preemption of events," is taken:

```
// This is the body method of the Queue class
public void body() {
  // assign a queue length
  int b = 10;
    while (Sim_system.running()) {
      //if number of customers do not exceed length b
      while (Sim_stat.QUEUE_LENGTH <= b) {

        Sim_event e = new Sim_event();
      // Accept events only from the source entity
      sim_get_next(e);
      sim_process(Sim_stat.THROUGHPUT);
      sim_completed(e);
      //continue to deal with the server
        sim_schedule(out1, 0.0, 1);
        sim_trace(1, "S Out1");
        //}
```

```
else {
    //No action! Just ignore events until statement above is
        true
    // then continue to accept events "customers"
    }
    } //end while statement
}}}...
```

9.12.3 A MATLAB Implementation of the M/M/1/b Queue

Observe how the size "capacity" of the queue is chosen and how to set it as a priority queue as an example of accepting customers according to their priorities. The three "sink" entities with their signal scopes are used to monitor the number of preempted customers. Notice that each customer has a priority value ("High," "Medium," or "Low"). A snapshot is shown in Figure 9.17.

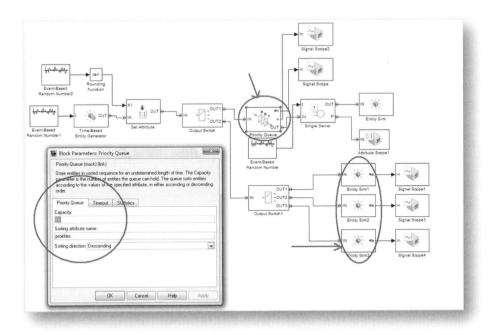

Figure 9.17   A snapshot of the MATLAB implementation of the *M/M/1/b* queue

9.13 The *M/M/m/m* Queue

This queue model is a further refinement of *M/M/1/b* that consists of a single queue with *m* servers and no storage capacity. Customers will arrive according to a Poisson process with rate λ. Each newly arriving customer is given their own private server. The probability distribution of the service time is exponential with mean $1/\mu$). This model is preferred for use in traditional telephony systems, where resources (circuits) are exclusive use. However, if a customer arrives when all *m* servers are busy, that customer is preempted. The number of servers that are busy in this queuing model can be easily illustrated as a BD process, as shown in Figure 9.18.

Analytically, the *M/M/m/m* queue exhibits the following properties:

- The probability that there are *n* customers in the system is:

$$P_n = P_0(\lambda/\mu)^n \frac{1}{n!} \quad n = 1, 2, 3, \ldots, m$$

such that:

$$P_0 = \left[\sum_{n=0}^{m} (\lambda/\mu)^n \frac{1}{n!} \right]^{-1}.$$

- The probability that an arriving customer faces full–busy servers and that customer is lost is:

$$P_m = \frac{(\lambda/\mu)^m/m!}{\sum_{n=0}^{m} (\lambda/\mu)^n/n!}.$$

- The serving throughput is:

$$\lambda \left[1 - \frac{\rho^m/m!}{\sum_{n=0}^{m} \rho^n/n!} \right]$$

where $\rho = \lambda/\mu$.

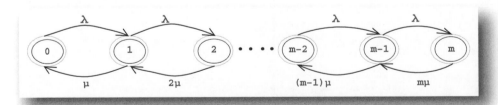

Figure 9.18   State transition diagram of the queuing model of the *M/M/m/m* queue

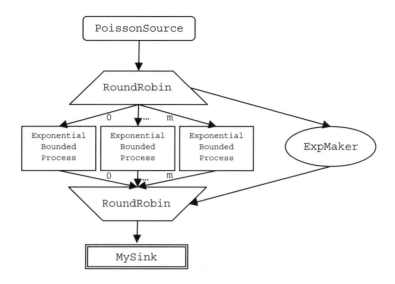

Figure 9.19 Flow chart of the CASiNO simulation architecture of the *M*/*M*/*m*/*m* queue

Within a more complex real-life simulation model, the *M*/*M*/*m*/*m* queue may frequently arise as a substructure. The above formal properties may then be used to validate the simulation results, and (in the case of steady-state inputs) can serve as justification for further model simplifications.

9.13.1 A CASiNO Implementation of the M/M/m/m Queue

Building on the classes designed in the case study of CASiNO in Chapter 8, the CASiNO code implementing an *M*/*M*/*m*/*m* queuing system is shown here. It consists of a PoissonSource which acts as source of packets, and two RoundRobin muxes around a factory in which ExpMaker creates ExponentialBoundedProcess objects dynamically, in response to the arrival of InitializeVisitors. The lower RoundRobin mux is connected to a MySink module where all visitors die.

The architecture of the CASiNO simulation flow chart is shown in Figure 9.19.

The code that creates this is:

```
public class Run_MMmm {

  public static void main (String [ ] args) {
    if (args.length != 4) {
      System.out.println ("Arguments: numpackets source-
        intensity "+" processing-exp numservers");
      System.exit (-1);
```

```
   }
   int numpackets = Integer.parseInt(args[0]);
   double source = Double.parseDouble(args[1]);
   double processing = Double.parseDouble(args[2]);
   double numservers = Double.parseDouble(args[3]);

   Conduit sc = new Conduit("poisson source",
                            new PoissonSource(numpackets,
                              source));
   Conduit rc1 = new Conduit("roundrobin-top",
     new RoundRobin());
   Conduit fc = new Conduit("exp-factory",
     new ExpMaker());
   Conduit rc2 = new Conduit("roundrobin-bottom",
     new RoundRobin());
   Conduit tc = new Conduit("sink", new MySink());
   Conduit.join(sc.getHalf(Conduit.ASIDE), rc1.getHalf
     (Conduit.ASIDE));
   Conduit.join(rc1.getHalf(Conduit.BSIDE), fc.getHalf
     (Conduit.ASIDE));
   Conduit.join(fc.getHalf(Conduit.BSIDE), rc2.getHalf
     (Conduit.BSIDE));
   Conduit.join(rc2.getHalf(Conduit.ASIDE), tc.getHalf
     (Conduit.ASIDE));

   for (int i=0; i<numservers; i++) {
     InitializeVisitor v =
       new InitializeVisitor(i, processing, true, 1);
     rc1.acceptVisitorFrom(v, Conduit.ASIDE);
   }

   Thread t = new Thread(Scheduler.instance());
   t.start();
   try { t.join(); }
   catch (Exception e) { }
   CASINO.shutdown();
   }
}
```

The program takes four arguments:

- the number of packets;
- the intensity of the Poisson process that governs the generation of packets in the source;
- the intensity of the exponential distribution that governs the packet processor's time to process each packet;
- the number of servers.

If we run this program once, specifying that 1000 packets are injected at a Poisson intensity of 2.0, with the intensity of the exponential processing delay being 10.0, we get the following results.

When the number of servers is 2:

```
SUMMARY bexp-0: fromA, total 168, max 1.0, mean=1.0, dropped
  333
SUMMARY bexp-1: fromA, total 163, max 1.0, mean=1.0, dropped
  338
SUMMARY poisson source: sent 1000 packets, received 0 packets
SUMMARY roundrobin-top: recvd 1000 packets
SUMMARY exp-factory: made 2 Exps
SUMMARY sink: received 331 packets
```

When the number of servers is 4:

```
SUMMARY bexp-0: fromA, total 146, max 1.0, mean=1.0, dropped
  105
SUMMARY bexp-1: fromA, total 160, max 1.0, mean=1.0, dropped
  91
SUMMARY bexp-2: fromA, total 148, max 1.0, mean=1.0, dropped
  103
SUMMARY bexp-3: fromA, total 147, max 1.0, mean=1.0, dropped
  104
SUMMARY poisson source: sent 1000 packets, received 0 packets
SUMMARY roundrobin-top: recvd 1000 packets
SUMMARY exp-factory: made 4 Exps
SUMMARY roundrobin-bottom: recvd 0 packets
SUMMARY sink: received 601 packets
```

When the number of servers is 8:

```
SUMMARY bexp-0: fromA, total 112, max 1.0, mean=1.0, dropped
  14
```

```
SUMMARY bexp-1: fromA, total 114, max 1.0, mean=1.0, dropped
  12
SUMMARY bexp-2: fromA, total 118, max 1.0, mean=1.0, dropped 8
SUMMARY bexp-3: fromA, total 114, max 1.0, mean=1.0, dropped
  12
SUMMARY bexp-4: fromA, total 118, max 1.0, mean=1.0, dropped 8
SUMMARY bexp-5: fromA, total 117, max 1.0, mean=1.0, dropped 9
SUMMARY bexp-6: fromA, total 117, max 1.0, mean=1.0, dropped 9
SUMMARY bexp-7: fromA, total 116, max 1.0, mean=1.0, dropped
  10
SUMMARY poisson source: sent 1000 packets, received 0 packets
SUMMARY roundrobin-top: recvd 1000 packets
SUMMARY exp-factory: made 8 Exps
SUMMARY sink: received 926 packets
```

These numbers can then be validated against the previously described theoretical predictions of the behavior of an *M/M/m/m* system.

9.13.2 A SimJava Implementation of the M/M/m/m Queue

Using SimJava, we create an *M/M/m/m* queue model with the required entities. The Queue class is no longer needed, so it was removed from the simulation running method leaving only one source and five servers ($m = 5$), and assigning events to each server equally according to the arrival rate λ. The ProcessorSubsystem main class is provided below with the required changes:

```
public class ProcessorSubsystem extends Anim_applet {
public void anim_layout() {

  //The source's delay between jobs is exponentially
    distributed
  Source source = new Source ("Source", 0, 0, "source", 1);

  Server server1 = new Server ("server1", 800, 20, "disk1",
    0.3);
  Server server2 = new Server ("server2", 800, 120, "disk1",
    0.3);
  Server Server3 = new Server ("server3", 800, 220, "disk1",
    0.1);
  Server Server4 = new Server ("server4", 800, 320, "disk1",
```

```
  0.1);
Server Server5 = new Server ("server5", 800, 420, "disk1",
  0.1);

//linking...
Sim_system.link_ports("Source", "Out1", "server1",
  "In");
Sim_system.link_ports("Source", "Out2", "server2",
  "In");
Sim_system.link_ports("Source", "Out3", "server3",
  "In");
Sim_system.link_ports("Source", "Out4", "server4",
  "In");
Sim_system.link_ports("Source", "Out5", "server5",
  "In");
}

public void sim_setup() {
  //finish after 10000 simulation time units
  Sim_system.set_transient_condition(Sim_system.TIME_
    ELAPSED, 1000);
  Sim_system.set_termination_condition(Sim_system.
    EVENTS_COMPLETED, "Processor", 0, 100, false);
  //make 2 replications and produce 95% confidence intervals
  Sim_system.set_output_analysis(Sim_system.IND_
    REPLICATIONS, 2, 0.95);
  Sim_system.set_report_detail(true, true);
  //publish a graph
  Sim_system.generate_graphs(true);
}... ... }}...
```

Finally, a snapshot of the simulation output is shown in Figure 9.20.

9.13.3 A MATLAB Implementation of the M/M/m/m Queue

We use "sim_event" in MATLAB to design an *M/M/m/m* queue model, creating the required entities and setting suitable attributes. One can see how to set the preemption property to each one of the five servers based on priority (the higher priority before the lower one). Then, the source directly connected to each of the servers. A snapshot of the MATLAB implementation of this queue is shown in Figure 9.21.

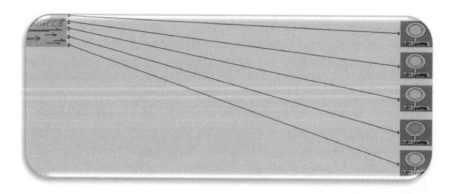

Figure 9.20   A snapshot of the SimJava implementation of the $M/M/m/m$ queue

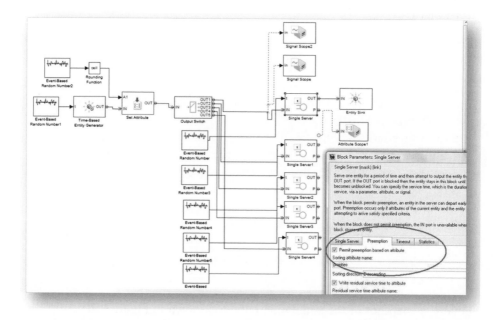

Figure 9.21   A snapshot of the MATLAB implementation of the $M/M/m/m$ queue

9.14 Summary

In this chapter, continuous and discrete random variables have been discussed. The use of queuing theory, state transitions matrices, state diagrams, and the birth–death process were discussed with elaborate examples. Some of the most commonly used queuing theory techniques have been presented and analyzed. Their corresponding code using CASiNO, SimJava, and MATLAB simulation techniques has been written and discussed.

Recommended Reading

[1] L. Kleinrock, *Queuing Systems, Volume I: Theory*, John Wiley & Sons, Inc., 1975.
[2] A. O. Allen, *Probability, Statistics, and Queuing Theory*, Academic Press, 1978.
[3] Online: http://en.wikipedia.org/wiki/Chapman-Kolmogorov_equation

10

Input Modeling
and Output Analysis

Input models greatly influence the outcomes of a simulation model. In real-world simulation applications, it is crucial that one chooses the appropriate distributions to represent the input data. If the distribution of inputs is misinterpreted, it will lead to inaccurate conclusions about the system. This chapter elaborates on the importance of data collection as a phase within the network modeling project life cycle. It lists the different data types that need to be collected to support network modeling projects, how to collect the data, choose the right distribution, and validate the correctness of one's choice.

Input models are used to represent the characteristics of randomness or uncertainty of the input source. *Input modeling* is the act of choosing the representation that best expresses the source data. This task often involves choosing an appropriate family of probability distributions. The task is simplified if one can make any of the following assumptions:

- The input data can be described as a sequence of independent random variables, and the distribution for each of them is identical.
- The common distribution shared by all the random variables is one of a general family of distributions that are widely implemented in most simulation systems or languages, e.g., the binomial, Poisson, normal, lognormal, exponential, gamma, beta, Erlang, Weibull, discrete or continuous uniform, triangular, or empirical distributions. The most commonly used of these are discussed in detail in Chapter 9.
- The input data can be represented as being generated by a stationary stochastic process, for which the probability distribution remains the same for all times. Its

Network Modeling and Simulation   M. Guizani, A. Rayes, B. Khan and A. Al-Fuqaha
© 2010 John Wiley & Sons, Ltd.

parameters, such as the mean and variance (if they exist), also do not change over time or position (see Chapter 9 for more details).

There are four major steps for developing a useful model of input data:

1. Collect data from a real-world scenario.
2. Choose a probability distribution to represent the input process.
3. Identify the parameters that are associated with the distribution family.
4. Evaluate the goodness of fit of the chosen distribution and the associated parameters.

These four steps are the standard procedure to follow when one wants a close match between the input model and the true underlying probability distribution associated with the system. Generally, there are also five basic questions that need to be answered when there is a given set of data being collected on the element of interest:

- Has the data been collected in an appropriate manner?
- Should a non-parametric model or a parametric probability distribution model be chosen as the input probabilistic mechanism?
- If a parametric model is chosen, what type of distribution will most adequately describe the data? In other words, what type of random variate generator seems to generate the same distribution of data?
- If a parametric model is chosen, what are the value(s) of parameter(s) that best state the probability distribution? If the distribution is *Gamma(a, b)*, for instance, then what is the value of *a* and *b*?
- After the probability distribution and its associated parameter(s) are chosen, how much confidence do we have that the chosen values are "best" to describe the input process?

Detailed discussions of these issues are presented in the following sections.

10.1 Data Collection

Data collection is the most important task when one tries to solve a real physical problem using a simulation model. Even if a good model structure is chosen, if it is poorly analyzed or inaccurate data is used, the results generated from the simulation will be meaningless. There are two approaches that can be elaborated with respect to the collection of data, the classical approach and the exploratory approach.

In the classical approach, a plan of collecting the data is followed. For instance, if one wants to study the probability distribution of inter-arrival times between packets arriving at a server, traffic flow could be recorded on a 24/7 basis in order to come up with the data set to analyze. This approach has the advantage that one knows exactly the data to be modeled. Researchers have proposed [2–5] some suggestions that might enhance the accuracy of the data being collected:

1. A well-planned and useful expenditure of time in the initial stages is important. The input process could have several forms, and it is very likely that the data collection process is modified several times before the actual data collection, the procedure becomes well designed. It is also necessary to know how to handle the unusual circumstances in order to ensure that the appropriate data is available.
2. Try to analyze and collect data simultaneously in order to rule out any data being collected that is useless to the simulation.
3. Try to associate homogeneous data sets. The quantile–quantile (Q–Q) plot can be performed for testing the homogeneity between two different data sets. For example, we might check for homogeneity of data during the same period of time on a Monday and Tuesday.
4. Try to find out whether there is a relationship between two variables, e.g., by plotting a scatter diagram to help determine whether two variables are related.
5. Beware that a sequence of independent observations actually may exhibit autocorrelation.

Some of the pitfalls that could occur during data collections are:

- False aggregation, e.g., model traffic flows per day, but only monthly data is available.
- False classification in time, e.g., data is available for this month, but it is necessary to model the traffic data for next month.
- False classification in space, e.g., it is desired to model the traffic flows at the arrival side, but only departure traffic data is recorded.

10.2 Identifying the Distribution

In this section, we discuss some strategies for selecting families of probability distributions which can best represent the input process after sample data is collected. This aspect of the process is more dependent on intuition and experience than it is on formal theory. For this reason, it is also considered the most difficult part of the overall process of fitting a distribution to input data. Fortunately, there are some questions that could help eliminate some probability models:

- What is the *source* of the data? Is the data finite? Is the data discrete or continuous? Is the data non-negative? Is it generated by a stationary arrival process?

- Plot a histogram and analyze its shape. This process will rule out several distributions from further consideration. Does the histogram follow some specific pattern of the probability distribution? Is it symmetric about the mean? The different histogram parameters can result in many different shapes of the histogram and one may plot several histograms with different parameters in order to acquire a better accuracy of data distribution to the hypothesis. The histogram can be constructed as follows:
  1. Break up the range of data into equal-width adjacent intervals.
  2. Based on the selected intervals, try to appropriately label the horizontal axis.
  3. Find out each interval's frequency of occurrence.
  4. Based on the frequencies of each interval, try to appropriately label the vertical axis.
  5. Plot the frequencies on the associated vertical axis.

Select the number of intervals approximately equal to the square root of the sample size. The width of the intervals should be also chosen properly. If this width is too wide, the shape of the histogram and details will not give good results. If it is too narrow, the histogram will be ragged and not smooth. Figure 10.1 presents these scenarios:

If the data is continuous, the shape of its histogram will be similar to the probability density function (PDF) of a theoretical distribution. On the other hand, if the data is discrete, the histogram associated with it will look like a probability mass function (PMF) of a theoretical probability distribution.

- Summarize some simple sample statistics in order to help filter the potential choices of the input data distribution. The following are useful summary statistics:
  (a) Minimum X_1 and maximum X_n values of data: They are rough estimates of the range.
  (b) Mean μ: This is the measure of central tendency, and the sample estimate is defined by:

$$\bar{X} = \frac{\sum_{i=1}^{n} X_i}{n}.$$

  (c) Median $x_{0.5}$: This is an alternative measure of the central tendency, and the sample estimate is defined by:

$$\hat{x}_{0.5} = \begin{cases} X_{(n+1)/2} & \text{if } n \text{ is odd} \\ \dfrac{[X_{(n/2)(n/2)} + X_{(n/2)+1}]}{2} & \text{if } n \text{ is even.} \end{cases}$$

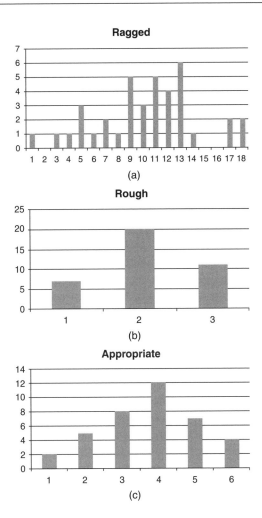

Figure 10.1 Data analysis: (a) ragged, (b) rough, and (c) appropriate

(d) Variance σ^2: This is the measure of variability, and the sample estimate is defined by:

$$s^2 = \frac{\sum_{i=1}^{n} X_i^2 - n\bar{X}^2}{n-1}.$$

(e) Lexis ratio τ: This is an alternative measure of variability for a discrete distribution. The Lexis ratio is very useful in distinguishing among the Poisson, binomial, and negative binomial distributions. If $\tau = 1$, this refers to the Poisson distribution; $\tau < 1$ refers to the binomial distribution; and $\tau > 1$ refers to the negative binomial distribution. The Lexis ratio statistic is given by:

$$\tau = \frac{S^2}{\bar{X}}.$$

(f) Coefficient of variation $cv = \sqrt{\sigma^2}/\mu$: This is an alternative measure of variability which can provide useful information on the form of a continuous distribution. For example, if the value of the coefficient of variation is close to 1, it suggests that the underlying probability distribution is exponential. The sample estimate cv is defined by:

$$\hat{cv} = \frac{\sqrt{S^2}}{\bar{X}}.$$

(g) Skewness $v = E[(X - \mu)^3]/(\sigma^2)^{3/2}$: This is a measure of symmetry. For instance, the potential input mode would be the normal or uniform distribution if the sample skewness is close to 0, and the sample estimate is defined by:

$$\hat{v} = \frac{\sum_{i=1}^{n} [X_i - \bar{X}]^3 / n}{(S^2)^{3/2}}.$$

10.3 Estimation of Parameters for Univariate Distributions

After one or more suitable families of distributions have been selected or hypothesized during the process as discussed in the previous section, one has to specify the value of their associated parameters in order to have a completely appropriate distribution specified for the input process. The *i.i.d.* data that has just been collected will be used to help choose the distribution which might be the best representation of the input process. In addition, the same data could be used to help estimate the parameters of the chosen distribution.

Calculation of the mean and variance of the collected data is a good estimation of the parameters of the distribution (the techniques for the computation of the mean and variance of the sample data were given in Section 7.1), if the sample raw data is either discrete or continuous. However, there is a situation that we might encounter where the sample data is discrete and has been grouped in a frequency distribution. Then the sample mean can be calculated as:

$$\bar{X} = \frac{\sum_{j=1}^{k} f_j X_j}{n}$$

and the sample variance as:

$$S^2 = \frac{\sum_{j=1}^{k} f_j X_j^2 - n\bar{X}^2}{n-1}$$

where k is the number of values of X and f_j is the frequency of X_j of X.

If the data is discrete or continuous and has been put into class intervals, then it is impossible to obtain the exact sample mean and variance. In this case, one can calculate the approximate value of the sample mean by:

$$\bar{X} \doteq \frac{\sum_{j=1}^{c} f_j m_j}{n}$$

and the sample variance is:

$$S^2 \doteq \frac{\sum_{j=1}^{c} f_j m_j^2 - n\bar{X}^2}{n-1}$$

where f_j is the observed frequency in the jth class interval, m_j is denoted as the midpoint of the jth interval, and c is the number of class intervals.

There are two approaches for estimating the parameters of the distribution from the sample data: the method of moments; and maximum likelihood estimation techniques. Both of them produce similar or identical results for many distribution estimations. The general setting for those techniques are the set of i.i.d. data denoted by $X_1, X_2, \ldots, X_n$ and one or more potential distributions that were chosen. Let q denote the number of unknown parameters (e.g., *Uniform(a, b)* model, $q = 2$).

The *method of moments* is an algebraic method for estimating the population parameters by equating the first q population moments to the first q sample moments given by:

$$E(X^k) = \frac{1}{n} \sum_{i=1}^{n} x_i^k.$$

For $k = 1, 2, \ldots, q$, and to solve the $q \times q$ set of equations for the unknown parameters q, where $E(X)$ is the expectation of a random variable X, or the first moment of the random variable, its variance, σ^2 or $V(X)$, can also be defined by:

$$\sigma^2 = E(X^2) - [E(X)]^2.$$

For instance, suppose we want to estimate the parameters μ and $\sigma^2 (q = 2)$ for a *normal* distribution; then we have to equate the first two population moments to the fist two sample moments:

$$E(X) = \frac{1}{n}\sum_{i=1}^{n} x_i \quad \text{and} \quad E(X^2) = \frac{1}{n}\sum_{i=1}^{n} x_i^2.$$

Let the above two equations be denoted by m_1 and m_2 respectively, and using the relationship $\sigma^2 = E(X^2) - [E(X)]^2$, the two equations can be rewritten as $\mu = m_1$ and $\sigma^2 = m_2 - \mu^2$. Therefore, by solving for μ and σ^2 one can compute the estimators of the normal distribution as:

$$\hat{\mu} = \bar{X} = m_1 \quad \text{and} \quad \hat{\sigma}^2 = \sigma^2 = m_2 - m_1^2.$$

That is to say, the method-of-moments technique estimates the mean and variance of the normal distribution by its sample mean and sample variance.

The maximum likelihood estimator (MLE) method is a technique that estimates the unknown parameters associated with either a hypothesized discrete distribution (PMF) or a continuous distribution (PDF) and provides an optimum fit.

Let us assume that there is a set of *i.i.d.* continuous random variables denoted by $X_1, X_2, \ldots, X_n$. If $\boldsymbol{\theta} = \{\theta_1, \theta_2, \ldots, \theta_q\}$ is a vector of unknown parameters of the hypothesized continuous distribution, then the likelihood function is given by:

$$L(\boldsymbol{\theta}) = \prod_{i=1}^{n} f(X_i, \boldsymbol{\theta}).$$

Note that the likelihood function $L(\boldsymbol{\theta})$ is the product of the PDF of each evaluated data value. Since the observations are independent, $L(\boldsymbol{\theta})$ can be rewritten as:

$$L(\boldsymbol{\theta}) = f_\theta(X_1) f_\theta(X_2) \ldots f_\theta(X_n).$$

As an example, let us assume that a set of *i.i.d.* data $X_1, X_2, \ldots, X_n$ is said to be exponentially distributed with an unknown parameter μ. By using maximum likelihood estimation, the likelihood function of the exponential distribution can be written as:

$$L(\mu) = \prod_{i=1}^{n} f(X_i, \mu) = \prod_{i=1}^{n} \frac{1}{\mu} e^{-X_i/\mu}$$

$$= \mu^{-n} \exp\left(-\frac{1}{\mu}\sum_{i=1}^{n} X_i\right).$$

To find the value of μ that maximizes $L(\mu)$, one can use the *log-likelihood function* defined as:

$$\ln L(\mu) = -n\ln\mu - \frac{1}{\mu}\sum_{i=1}^{n}X_i.$$

In order to maximize the log-likelihood function, standard differential calculus can be used with respect to μ to obtain:

$$\frac{\partial \ln L(\mu)}{\partial \mu} = -\frac{n}{\mu} + \frac{1}{\mu^2}\sum_{i=1}^{n}X_i.$$

Equating to zero and solving for μ, we can obtain the maximum likelihood estimator as:

$$\hat{\mu} = \frac{1}{n}\sum_{i=1}^{n}X_i.$$

That is, the sample mean $\bar{X}$ of the data. It is clear that the method of moments and maximum likelihood estimator generate identical results.

Now we use the same approach to estimate the parameters of discrete distributions using likelihood function $L(\boldsymbol{\theta})$ as:

$$L(\boldsymbol{\theta}) = \prod_{i=1}^{n}p(X_i, \boldsymbol{\theta}) = p_\theta(X_1)p_\theta(X_2)\ldots p_\theta(X_n)$$

where $p_\theta(x)$ is the PMF for the chosen distribution with unknown parameter θ.

For instance, we assume that there is a given data set $X_1, X_2, \ldots, X_n$ that is geometrically distributed and $p(x) = p(1-p)^x$ for $x = 0, 1, \ldots$. Then, the likelihood function of the geometric distribution is:

$$L(p) = p^n(1-p)^{\sum_{i=1}^{n}X_i}.$$

By taking logarithms, we obtain:

$$\ln L(p) = n\ln p + \sum_{i=1}^{n}X_i\ln(1-p).$$

Then, differentiating $L(p)$ we get:

$$\frac{\partial \ln L(p)}{\partial p} = \frac{n}{p} - \frac{\sum_{i=1}^{n}X_i}{1-p}.$$

Table 10.1 Suggested estimators for common distributions

Distribution	Parameter(s)	Suggested estimator(s)
Poisson	λ	$\hat{\lambda} = \bar{X}$
Exponential	λ	$\hat{\lambda} = 1/\bar{X}$
Gamma	β, θ	$\hat{\beta}, \hat{\theta} = 1/\bar{X}$
Normal	μ, σ^2	$\hat{\mu} = \bar{X}, \hat{\sigma}^2 = S^2$
Lognormal	μ, σ^2	$\hat{\mu} = \bar{X}, \hat{\sigma}^2 = S^2$

Equating this to zero and solving for p, we obtain the maximum likelihood estimator of the geometric distribution as:

$$\hat{p} = 1/(\bar{X} + 1).$$

Both the method of moments and the maximum likelihood estimator are very useful techniques for finding the best parameter values of the hypothesized distribution. But, in some respects, the method of moments has a poorer reputation than the maximum likelihood estimator when estimating parameters of a known family of probability distributions. This is because the maximum likelihood estimators have a higher probability of being close to the quantities to be estimated.

However, in some circumstances, such as the gamma distribution, the equation of the likelihood function may be difficult to use without a calculating machine, such as a computer. On the other hand, the method-of-moments estimators can be easily calculated by hand.

Table 10.1 gives the estimators that are used by researchers for such calculations.

10.4 Goodness-of-Fit Tests

After selecting one or more probability distributions that might best represent the sample data, one could go deeper and carefully examine these distributions to see the discrepancy between the collected sample data and the expected "quality" of the representation for the data of the chosen distribution. If there is more than one candidate for the distribution, then a decision should be made on the one that can provide the best fit for the input process. The final goal of this process is to determine a probability distribution that is accurate enough for the intended purposes of input modeling.

There are two approaches to testing for goodness of fit: a graphical approach and an analytical approach.

There are two types in the graphical comparison approach: "P–P" (probability–probability) plot and "Q–Q" (quantile–quantile) plot. These are widely and commonly used in testing the goodness of fit. The P–P plot tests whether the sample data follows a given distribution. One can visually test the location and scale parameters that are associated with the chosen distribution. This can be thought of as a graphical comparison of an estimated CDF of the collected data to the fitted CDF of the distribution. In other words, it is the fitted CDF at the ith order statistic X_i, $F_n(X_{(i)}) = i/n$, versus the empirical CDF $\tilde{F}_n(X_{(i)}) = F_n(X_{(i)}) - 0.5/n = i - 0.5/n$, for $i = 1, 2, \ldots, n$. The good fit of the chosen distribution can be indicated if the points of the plot form approximately a straight line.

The second graphical approach, the Q–Q plot, is a useful tool for evaluating the distribution fit. It is also a graphical comparison of the sample data to the quantile value of the CDF of the chosen distribution. Let $q = (1 - 0.5)/n$, so that $0 < q < 1$,; then the q quantile of the random variable X is the value γ such that $F(\gamma) = q$. Because the construction of the Q–Q plot needs the calculation of the distribution quantile, the inverse transformation of CDF F, then we write $\gamma = F^{-1}(q)$. Therefore, the Q–Q plot is the comparison of each sorted sample data versus its associated inverse CDF of the distribution. For instance, assuming there is a random variable X with CDF F, let $\{x_i, i = 1, 2, \ldots, n\}$ be the data from X; then data is sorted in ascending order, denoted by $\{k_j, j = 1, 2, \ldots, n\}$, where $k_1 \leq k_2 \leq \cdots \leq k_n$. Thus, $j = 1$ represents the smallest value and $j = n$ represents the largest value. If the chosen distribution is an appropriate distribution of the random value of X, then the value of k_j is approximately the value of $F^{-1}((j - 0.5)/n)$. Therefore, a plot of k_j versus $F^{-1}((j - 0.5)/n)$ will form approximately a straight line. On the other hand, if the points of the plot deviate from a straight line, this means that the hypothesized distribution is inappropriate. Figure 10.2 shows the Q–Q plot for the exponential distribution with inter-arrival time data.

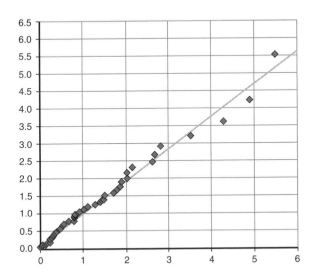

Figure 10.2   Q–Q plot for the exponential distribution quantile with associated data

Note that both P–P and Q–Q plots are subjective because different experts would have different visual preferences. One could consider Figure 10.2 as an example, while others might feel that it is nothing but a straight line. Thus, more objective techniques have to be adopted. These are the so-called analytical approach techniques.

In the analytical approach, there are two standard statistical techniques that can be used: the Kolmogorov–Smirnov test and the chi-square test.

10.4.1 Chi-Square Goodness-of-Fit Test

The chi-square (χ^2) test is a technique that tests for a null hypothesis that the distribution of a certain observed event's frequency in a sample is consistent with the hypothesized discrete or continuous distribution. In addition, the size of the sample data must be large enough to make the chi-square goodness-of-fit test more accurate when the associated parameters are estimated by maximum likelihood estimators.

The first step in the chi-square test is to arrange the number of observations n into a set of k cells (class intervals), and then calculate the chi-square statistic by finding the difference between each observed and expected frequency. The statistical formula of the chi-square test is given by:

$$\chi_0^2 = \sum_{i=1}^{k} \frac{(O_i - E_i)^2}{E_i}$$

where O_i is the observed frequency in the ith cell and E_i is the expected frequency in the same cell, which is asserted by the null hypothesis, and the expected frequency can be computed by np_i, where p_i is the hypothesized probability of the ith cell.

Another element of the chi-square test is the degrees of freedom. These can be computed using $k - s - 1$, where s denotes the number of parameters associated with the hypothesized distribution. The hypotheses of the chi-square test are either true or false, depending on its critical value with a certain value of level of significance $x_{\alpha, k-s-1}^2$. The null hypothesis will be rejected if χ_0^2 is bigger than the critical value.

When applying the chi-square goodness-of-fit test, the most difficult problems that one might encounter are choosing the number of cells and finding the minimum value of the expected frequency. In fact, there is no good definition that can generate and guarantee good results. Fortunately, there are a few guidelines that can be followed to achieve this.

First, let us consider the continuous distribution scenario. If the distribution being tested is continuous, then a technique called the *equiprobable approach* can be used to

Table 10.2 Possible number of class intervals for continuous data

Sample size n	Number of class intervals k
20	Too small for chi-square test
50	5 to 10
100	10 to 20
>100	$\sqrt{n}$ to $n/5$

divide the data into several cells or class intervals. Each of these cells has the same probability, so that $p_1 = p_2 = \cdots = p_i$, and can be defined as:

$$p_i = \int_{a_{i-1}}^{a_i} f(x)dx = F(a_i) - F(a_{i-1})$$

where a_i and a_{i-1} are the endpoints of the ith cell. Even though this might be inconvenient because the CDF of the fitted distribution has to be inverted, this technique is still the best approach for increasing the accuracy of the chi-square goodness-of-fit test. Table 10.2 gives the suggested number of class intervals for continuous data.

On the other hand, for the discrete scenario, each value of the random sample data should generally be a class interval or cell, unless it has to be combined with adjacent class intervals in order to satisfy the minimum expected cell frequency requirement. If combining adjacent cells is not necessary, then $p_i = p(x_i) = P(X = x_i)$.

Unfortunately, there is no solution for finding the probability associated with each cell that maximizes the power (probability of rejecting the false hypothesis) of the chi-square test of a given data size. However, the chi-square test can be effective if $k \geq 3$ and the minimum value of expected frequency is greater than or equal to 5. Therefore, if an expected frequency is too small, it can be combined with the expected frequency of an adjacent cell, and the corresponding observed frequency should also be combined. Then the value of k should be modified depending on how many cells are combined.

Based on the properties mentioned above, if the equiprobable approach is used, then $p_i = 1/k$, $E_i = np_i \geq 5$, so substituting for p_i to obtain $n/k \geq 5$, and solving for k gives $K \leq n/5$.

10.4.2 Kolomogorov–Smirnov Goodness-of-Fit Test

From the previous section, we learned that the chi-square goodness-of-fit test requires data to be divided into class intervals. In the continuous scenario, this process is unreasonable and sometimes will cause possible loss of information. Changing the number of class intervals will make the chi-square test more arbitrary, since a hypothesis could be rejected when data is arranged in one way, but accepted when it is arranged in another way. As a result, the Kolomogorov–Smirnov goodness-of-fit test has been more widely accepted than the chi-square goodness-of-fit test.

The Kolomogorov–Smirnov (K-S) test formalizes the idea behind the Q–Q plot. It is a goodness-of-fit test used to determine whether an underlying probability distribution differs from a hypothesized distribution when given a finite data set. It compares an empirical distribution function with the CDF F that is specified by the hypothesized distribution.

Compared to the chi-square test, there are some advantages of the K–S test that we should note. First, K–S tests do not require data to be grouped into class intervals, so it eliminates the difficulty of determining the appropriate number of intervals. Second, the K–S test is suitable for any sample size when no parameters are estimated from the data, whereas chi-square tests are valid only for large data sets.

The K–S goodness-of-fit test is simply the largest vertical distance, denoted by D_n, between the empirical distribution function and the fitted CDF of the chosen distribution. So, the statistic of the K–S test can be defined by:

$$D_n = \sup_x \{|F_n(x) - \hat{F}(x)|\}$$

where "sup" is the *supremum* of a set, i.e., it is the least (smallest) element that is greater than or equal to all elements in that set. Thus, based on this equation, the statistic D_n can be computed as:

$$D_n^+ = \max_{1 \le i \le n} \left\{ \frac{i}{n} - \hat{F}(X_{(i)}) \right\}, \quad D_n^- = \max_{1 \le i \le n} \left\{ \hat{F}(X_{(i)}) - \frac{i-1}{n} \right\}$$

and then we let $D_n = \max\{D_n^+, D_n^-\}$. After computing the value of D_n, we use it to compare the critical value to the associated level of significance value. If D_n is greater than the critical value, it means that the hypothesis of the distribution must be rejected.

In order to increase the flexibility of the K–S test, this technique has been modified so that it can be used to deal with many situations where the parameters are estimated from the data. Different tables of critical values are used for dealing with different situations of distribution hypotheses. Recently, standard tables of K–S tests were modified in terms of the critical values to calculate the goodness of fit for normal and exponential distributions.

Another similar technique to the K–S test is the *Anderson–Darling test* (A–D test). This is implemented based on the properties of the K–S test. The A–D test also compares the difference between the empirical CDF and fitted distribution CDF, like the K–S test. But, unlike the K–S test, the A–D test is more focused on the measure of difference, not just the maximum difference, and it is more sensitive to the discrepancies in the tails of the distribution, which makes the A–D test more powerful than the K–S test against many alternative distributions. The A–D statistic A_n^2 is given by:

$$A_n^2 = n \int_{-\infty}^{\infty} [F_n(x) - \hat{F}(x)]^2 \psi(x) \hat{f}(x) dx$$

where the *weight function* $\psi(x) = 1/[\hat{F}(x)(1 - \hat{F}(x))]$. Therefore, A_n^2 is the weighted average of squared differences between the empirical CDF and fitted CDF $[F_n(x) - \hat{F}(x)]^2$, and the largest weight for $\hat{F}(x)$ is close to 1 (right tail) and close to 0 (left tail). So, if we let $Z_i = \hat{F}(X_{(i)})$ for $i = 1, 2, \ldots, n$, then the above equation can be rewritten as:

$$A_n^2 = \left(-\left\{ \sum_{i=1}^{n} (2i - 1)[\ln Z_i + \ln(1 - Z_{n+1-i})] \right\} / n \right) - n$$

which is the actual calculation of the statistic. Thus, the null hypothesis H_0 will be rejected if A_n^2 exceeds the critical value $a_{n,1-\alpha}$, where α is the level of significance of the test.

10.5 Multivariate Distributions

So far, we have just considered only the estimation of the single, univariate distribution of the random variable at a time. However, there are some systems in which input data is statistically related to other data. For instance, several random input variables together may form a random vector with some joint (multivariate) probability distribution, or there could be a correlation between different input random variables; in this case, we have to specify the correlation between those input random variables in order to increase our model's validity. In short, if we have discovered some kind of statistical relationship between our input random variables, or the input process is a process over time that shows the autocorrelation within itself, then we must pay extra attention to modeling those properties and relationships in our input modeling process.

10.5.1 Correlation and Covariance

In a real situation, it is often the case that two random variables are *correlated*, that is to say, the natural property among these two variables is coupled, which indicates the strength and direction of a linear relationship between these two random variables. Thus, the measurements of covariance and correlation are an indication of how good the relationship is between two random variables, say X_i and X_j. Let $\mu_i = E(X_i)$ and $\sigma_i^2 = V(X_i)$ be the mean and variance of X_i, respectively. Then the measurement of covariance is an indication of the linear dependence between X_i and X_j, where

$i = 1, 2, \ldots, n$ and $j = 1, 2, \ldots, n$, which is denoted as C_{ij} or $\mathrm{cov}(X_i, X_j)$, and the calculation is defined by:

$$C_{ij} = E[(X_i - \mu_i)(X_j - \mu_j)] = E(X_i X_j) - \mu_i \mu_j.$$

Note that covariances are symmetric, which means $C_{ij} = C_{ji}$, and if $i = j$, then $C_{ij} = C_{ii} = \sigma_i^2$. Thus, the random variables X_i and X_j are said to be uncorrelated if C_{ij} is equal to zero.

We have two definitions that are associated with the covariance C_{ij}, namely positive correlation and negative correlation. If X_i and X_j are positively correlated, that means $C_{ij} > 0$, in other words, $X_i > \mu_i$ and $X_j > \mu_j$ are likely to occur together, and $X_i < \mu_i$ and $X_j < \mu_j$ are likely to occur together also. Thus, if the random variables are positively correlated, one is large and the other is also large. If X_i and X_j are negatively correlated, it means that $C_{ij} < 0$. In this situation, $X_i > \mu_i$ and $X_j < \mu_j$ are likely to occur together, while $X_i < \mu_i$ and $X_j > \mu_j$ are also likely to occur together; that is to say, if one is large, the other one is likely to be small when they are negatively correlated.

Thus, based on the relationship we have discussed above, the correlation (ρ) between two random variables can be defined by:

$$\rho = \mathrm{corr}(X_i, X_j) = \frac{C_{ij}}{\sqrt{\sigma_i^2 \sigma_j^2}} = \frac{C_{ij}}{\sigma_i \sigma_j}.$$

Let us consider a scenario where we want to evaluate the correlation for s sequences of consecutive i.i.d. random variables $X_1, X_2, X_3, \ldots$; it is often the case that we are interested in seeing if our set of data is autocorrelated. In this case, we pick a small, fixed positive integer k, which is called the *autocorrelation lag*, to calculate the serial correlation for a range of lag values.

Assume we have n consecutive random variables and we want to define the sample autocovariance for lag k, which can be computed as:

$$C_k = \frac{1}{n-k} \sum_{i=1}^{n-k} (X_i - \bar{X})(X_{i+k} - \bar{X}), \quad \text{for } k = 1, 2, \ldots$$

Thus, the associated autocorrelation for lag k is given by:

$$r_k = \frac{C_k}{C_0}$$

where $C_0 = S^2$ is the sample variance.

10.5.2 Multivariate Distribution Models

We have discussed how to find the correlation between two random input variables, so now let us consider how to specify the multivariate distributions.

Let X_1, X_2 to be normally distributed variables; then we can model their dependence by using a bivariate normal distribution with parameters μ_1, σ_1^2 and μ_2, σ_2^2, and then $\rho = \mathrm{corr}(X_1, X_2)$. Suppose that we have n distributed i.i.d. pairs $(X_{11}, X_{21}), (X_{12}, X_{22}), \ldots, (X_{1n}, X_{2n})$, so in order to estimate ρ, the sample covariance is defined as:

$$\hat{C}_{12} = \frac{1}{n-1} \sum_{i=1}^{n} (X_{1i} - \bar{X}_1)(X_{2i} - \bar{X}_2) = \frac{1}{n-1} \left(\sum_{i=1}^{n} X_{1i} X_{2i} - n \bar{X}_1 \bar{X}_2 \right)$$

where $\bar{X}_1$ and $\bar{X}_2$ are sample means of the data. Thus, the estimated correlation is defined by:

$$\hat{\rho} = \frac{\hat{C}_{12}}{\hat{\sigma}_1 \hat{\sigma}_2}$$

where $\hat{\sigma}_1, \hat{\sigma}_2$ are the sample variances.

Therefore, we can conclude that if we want to generate bivariate normal random variables, then we need to: first, generate independent, standard, normally distributed random variables D_1 and D_2; second, set $X_1 = \mu_1 + \sigma_1 D_1$; and third, set $X_2 = \mu_2 + \sigma_2(\rho D_1 + \sqrt{1 - \rho^2} D_2)$.

10.5.3 Time-Series Distribution Models

Let $X_1, X_2, \ldots$ be a sequence of random variables that are identically distributed, but are dependent on each other with stationary covariance; in order to understand how to model this type of input process, we will discuss two models that have autocorrelations like:

$$\rho_j = \mathrm{corr}(X_i, X_{i+j}) = \rho^j$$

for $j = 1, 2, 3, \ldots$.

First, we consider the time-series model called the stationary AR(1) (*autoregressive order 1*) model with mean μ:

$$X_i = \mu + \phi(X_{i-1} - \mu) + \varepsilon_i \quad i = 2, 3, \ldots$$

where the ε_i's are independent and identically normally distributed random variables with mean and variance 0 and σ_ε^2 chosen to control $V(X_i)$, and $-1 < \phi < 1$. The estimation of ϕ can be obtained by using the lag 1 autocorrelation:

$$\phi = \rho^1 = \text{corr}(X_i, X_{i+1}).$$

Thus, in order to estimate ϕ, we must first compute the estimated lag 1 auto-covariance by:

$$C_{i,i+1} = \frac{1}{n-1}\sum_{i-1}^{n-1}(X_i - \bar{X})(X_{i+1} - \bar{X}) \doteq \frac{1}{n-1}\left(\sum_{i-1}^{n-1}X_iX_{i+1} - (n-1)\bar{X}^2\right)$$

and we calculate the estimated value of ϕ by using the estimated variance:

$$\hat{\phi} = \frac{c_{i,i+1}}{\hat{\sigma}^2}.$$

Last, we estimate μ and σ_ε^2 from:

$$\hat{\mu} = \bar{X} \quad \text{and} \quad \hat{\sigma}_\varepsilon^2 = \hat{\sigma}^2(1 - \hat{\phi}^2).$$

Therefore, we can conclude that, given values of parameters ϕ, μ, and σ_ε^2, we can generate a stationary AR(1) time-series model by generating X_1 from a normal distribution with mean and variance μ and $\sigma_\varepsilon^2/(1 - \phi^2)$ respectively, and set $i = 2$; then we generate ε_i from the normal distribution with mean 0 and variance σ_ε^2; after that we set $X_i = \mu + \phi(X_{i-1} - \mu) + \varepsilon_i$. Finally, we let $i = i + 1$ and move to the second process.

Next, consider the time-series model called EAR(1) (*exponential autoregressive order 1*) model:

$$X_i = \begin{cases} \phi X_{i-1}, & \text{with probability } \phi \\ \phi X_{i-1} + \varepsilon_i, & \text{with probability } 1 - \phi \end{cases}$$

for $i = 2, 3, \ldots$, where the ε_i's are independent and identically exponentially distributed random variables with mean $1/\lambda$ and $-1 < \phi < 1$.

The estimation of parameters of EAR(1) is very similar to AR(1). The estimated lag 1 autocorrelation can be computed by setting $\hat{\lambda} = 1/\bar{X}$; then we get:

$$C_{i,i+1} = \frac{1}{n-1}\sum_{i=1}^{n-1}\left(X_i - \frac{1}{\bar{X}}\right)\left(X_{i+1} - \frac{1}{\bar{X}}\right) \doteq \frac{1}{n-1}\left(\sum_{i=1}^{n-1}X_iX_{i+1} - (n-1)\frac{1}{\bar{X}}\right)^2.$$

Then, we calculate the estimated value of ϕ by setting $\hat{\phi} = \hat{\rho}$, and obtain:

$$\hat{\phi} = \hat{\rho} = \frac{c_{i,i+1}}{\hat{\sigma}^2}.$$

Thus, we can conclude that, given values of the parameters ϕ and λ, we can generate a stationary EAR(1) time-series model by generating X_1 from an exponential distribution model with mean $1/\lambda$, and set $t = 2$; then we generate Y from a uniform distribution model with parameters 0 and 1; if $Y \leq \phi$, then we set $X_i = \phi X_{i-1}$, or else we can generate ε_i from an exponential distribution with mean $1/\lambda$ and let $X_i = \phi X_{i-1} + \varepsilon_i$. Finally, we let $i = i + 1$ and then move to the second process.

10.6 Selecting Distributions without Data

In a real-world scenario, it is often the case that it is not possible to collect data for our input modeling, so the techniques that we have mentioned in the previous section are not suitable for the problem of choosing appropriate probability distributions. For instance, if the system or scenario we want to simulate is still in the design process and does not yet exist in concrete form, then it is obvious that collecting data is impossible for this simulation. How can one model traffic at a gas station before any gas stations have been built?

Therefore, in such a situation, we must be well prepared and resourceful in selecting the input process model. In some circumstances, we can obtain information on the process in the absence of data by analyzing the nature of the process, such as inter-arrival times between customers at a specific store. If the customers arrive one at a time, then we might think that the inter-arrival times between each customer are exponentially distributed; or we could make some educated guesses based on expert opinion by talking to people who are experienced in the particular process, because they can provide intuition on which the choice of input distribution can be based.

The uniform, beta, and triangular distributions are often used as the input distribution when data is unavailable because of their characteristics. The uniform distribution can be a bad choice, because its upper and lower bounds seldom fit the central values in the real process, so in this case, if the upper and lower bounds are known, the triangular distribution can be used and the most likely value can be used as the peak of the distribution.

10.7 Output Analysis

Many simulation studies include randomness in order to get a sense of typical inputs that the system might face. For example, in the simulation of a processor system, the processing times required may follow a given probability distribution, or the arrival times of new jobs may be stochastically defined. Likewise, in a bank simulation, customers might arrive at random times and the amount of time spent at a teller could be stochastically determined. Because of the randomness in the components driving a simulation, its output is also random, so statistical techniques must be used to analyze the results and aggregate them into meaningful measures from which to draw conclusions.

The data analysis methods in introductory statistics courses typically assume that the data is i.i.d. with a normal distribution. Unfortunately, the output data from simulations is often not i.i.d. and not normal. For example, consider customer waiting times before seeing a teller in a bank. If one customer has an unusually long waiting time, then the next customer probably also will, so the waiting times of the two customers are dependent. Moreover, customers arriving during lunch hour will usually have longer waiting times than customers coming in at other times, so waiting times are not identically distributed throughout the day. Finally, waiting times are always positive and often skewed to the right, with a possible mode at zero, so waiting times are not normally distributed. For these reasons one often cannot analyze simulation output using the classical statistical techniques developed for i.i.d. normal data. For these reasons, more sophisticated statistical methods are needed to examine and analyze simulation experiments to measure performance.

One of the first steps in any simulation study is choosing the performance measure to calculate. In other words, what measures will be used to evaluate how "good" the system is. For example, the performance of a queuing system may be measured by the expected number of customers served in a day, or we may use the long-run average daily cost, or the inverse of the maximum queue size, or the inverse of the mean queue size as the measure of performance of a supply chain. There are primarily two types of performance measures for stochastic systems:

1. Transient performance measures, also known as terminating or finite-horizon measures, which evaluate the system's evolution over a finite time horizon.
2. Steady-state performance measures, which describe how the system evolves over an infinite time horizon. These are also known as long-run or infinite-horizon measures.

10.7.1 Transient Analysis

Transient analysis is an option when we are dealing with a terminating simulation in which there is a "natural" event B that specifies the length of time for which one is interested in the system. The event B often occurs either at a time point beyond which no useful information is obtained, or when the system is "cleaned out." For example, if we are interested in the performance of a system during the first 10 time units of operation in a day, then B would denote the event that 10 time units of system time have elapsed. If we want to determine the first time at which a queue has at least eight customers, then B is the event of the first time that the queue length reaches eight. Since we are interested in the behavior of the system over only a finite time horizon, the "initial conditions" I can have a large impact on the performance measure. For example, queuing simulations often start with no customers present, which would be I in this setting. In a transient simulation, the goal is to calculate:

$$\mu = E(X)$$

where X is a random variable representing the (random) performance of the system over some finite horizon.

Example 1:   Consider a bank line containing an automatic teller machine (ATM). The line is only open during normal banking business hours, which is 9.00 a.m. to 5.00 p.m., so a customer can access the ATM only during those times. Any customers in the line at 5.00 p.m. will be allowed to complete their transactions, but no new customers will be allowed in. Let Z be the number of customers using the ATM in a day. We are interested in determining the following terminating performance measures:

- $E(Z)$, the expected value of Z. To put things in the framework of (I), we set $X = Z$.
- $P\{Z \geq 500\} = E[I(Z \geq 500)]$, which is the probability that at least 500 customers use the ATM in a day, where $I(A)$ is the indicator function of an event A, which takes the value 1 if A occurs, and 0 otherwise. $X = I\ (Z \geq 500)$ in this case.

The initial conditions I might be that the system starts out empty each day, and the terminating event B is that it is past 5.00 p.m. and there are no more customers in the line.

Alternatively, we might define Z to be the average waiting time (in seconds) of the first 50 customers in a day. We can then define the following performance measures:

- $E(Z)$, the expected value of Z. In this case, $X = Z$.
- $P\{Z \leq 30\} = E[I(Z \leq 30)]$, which is the probability that the average waiting time of the first 50 customers is no more than 30 seconds. Here, $X = I(Z \leq 30)$.

In this case we might specify the initial conditions I to be that the system starts out empty each day, and the terminating event B is that 50 customers have finished their waits (if any) in line.

10.7.2 Steady-State Analysis

Now we consider steady-state performance measures. Let $Y = (y_1, y_2, y_3, \ldots)$ be a discrete-time process representing the output of a simulation. For example, if the booth containing the ATM in the previous example is now open 24 hours a day, then y_i might represent the waiting time of the ith customer since the ATM was installed.

Let $Fi(Y|I) = P(Yi \leq y|I)$ for $i = 1, 2, \ldots$, where, as before, I represents the initial conditions of the system at time 0.

Now, if $Fi(Y|I) \rightarrow F(y) \rightarrow$ as $i \rightarrow \infty$, then $F(y)$ is called the steady-state distribution of the process Y.

Many systems do not have steady-state characteristics. For example, consider our previous example of an ATM that is accessible only during business hours. Let Y_i be the waiting time of the ith customer to arrive since the ATM was installed. Then, the process Y does not have a steady state because the first customer of each day always has no wait, whereas other customers may have to wait. For example, suppose 500 customers are served on the first day, so the second day begins with customer 501, who has no wait since there is no one ahead in the line on that day. On the other hand, if the ATM were accessible 24 hours a day, then a steady state might exist.

Consider the ATM discussed earlier, but now suppose that it is accessible all the time. Let Y_i be the number of customers served on the ith day of operation, and suppose that, over time, the system goes into steady state. We now may be interested in determining the following steady-state performance measures:

- $E(Y)$, which is the expected steady-state number of customers served in a day.
- $P\{Y \geq 400\} = E[I(Y \geq 400)]$, which is the steady-state probability that at least 400 customers are served per day.

Again, we may let the initial conditions I denote that the system begins operations on the first day with no customers present, and, over time, the effects of the initial conditions "wash away."

10.8 Summary

This chapter discussed how to deal with data collection as a phase within the network modeling project life cycle. It lists the different data types that need to be collected to support network modeling projects, how to collect the data, choose the right distribution, and validate the correctness of the choice for the kind of distribution that represents such data.

Recommended Reading

[1] J. Banks, J. S. CarsonII, B. L. Nelson, and D. M. Nicol, *Discrete-Event System Simulation*, 4th Edition, Pearson Education, 2005.

[2] O. Balci, "Verification, validation, and certification of modeling and simulation applications," *Proceedings of the 2003 ACM Winter Simulation Conference* (S. Chick, P. J. Sanchez, D. Ferrin, and D. J. Morrice, eds.), ACM, 2003, pp. 150–158.

[3] A. M. Law, *Simulation Modeling & Analysis*, 4th Edition, McGraw-Hill, 2006.

[4] L. M. Leemis and S. K. Park, *Discrete-Event Simulation: A First Course*, Prentice Hall, 2006.

[5] B. L. Nelsonnd M. Yamnitsky, "Input modeling tools for complex problems," *Proceedings of the 1998 IEEE Winter Simulation Conference* (D. Medeiros, E. Watson, J. Carson, and M. Manivannan, eds.), IEEE, 1998, pp. 105–112.

11

Modeling Network Traffic

11.1 Introduction

As networks are becoming more pervasive, network designers and researchers are progressively focusing on optimizing network performance and improving network utilization in terms of throughput, end-to-end delay, packet loss, and other SLA-related parameters. Network performance and optimization cannot be realized without a full understanding of network traffic modeling together with traffic patterns and traffic generation. In order to generate realistic traffic patterns, mathematical models should accurately represent the relevant statistical properties of the original traffic.

Network traffic modeling is essential for theoretical capacity and performance representation as well as for simulation models. Traffic theory suggests using the application of mathematical modeling to explain the relationship between traffic demand and experimental performance. While network traffic models are well understood for classical networks (e.g., standard assumptions of Poisson arrival rates and exponential call holding time for the public switched telephone network (PSTN)), they are more complex for modern high-speed networks where traffic behavior is difficult to predict. Modern networks exhibit a mixture of traffic types including voice, data, and video, as well as mobility where the location of nodes and consequently the source of the traffic are not fixed.

In this chapter, we present several traffic models that are used by simulation researchers and practitioners to model network traffic loads. It also describes the most commonly used global optimization techniques to solve optimization problems. These optimization techniques are useful in modeling and solving complex physical problems.

Network Modeling and Simulation   M. Guizani, A. Rayes, B. Khan and A. Al-Fuqaha
© 2010 John Wiley & Sons, Ltd.

11.2 Network Traffic Models

Every few years, the amount of processing speed that one can buy doubles, and the same holds for memory chips as well. This has been the trend for several decades and it is expected to keep going for a few decades to come. This encourages customers to obtain more powerful computers with multimedia capabilities. Consequently, customers have begun to migrate from text to images, animation, and lately to video, acquiring high-speed networks capable of supporting new types of services. So the emergence of new types of services like Cisco's TelePresence, video conferencing, video on demand, video telephony, video library, high-definition television (HDTV), and other services with unknown requirements is expected to be phenomenal in the future.

Traffic models can be generally divided into two main categories: models which exhibit long-range dependencies or self-similarities; and Markov models that exhibit only short-range dependence (Markov chains are discussed in detail in Chapter 9). Self-similar models include the fractional Brownian motion (FBM) model, on/off models with heavy-tailed distributions for the on/off duration, and the M/Pareto/Infiniti models.

Markov models include the traditional on/off models with exponential on/off distributions, the Markov-modulated Poisson process, and Gaussian autoregressive models, which typically have exponentially decaying correlation functions.

In the following, we study two traffic models that fall into the second category, namely constant bit rate and variable bit rate. We also consider one traffic source that falls into the second category, namely the Pareto traffic (self-similar traffic).

11.2.1 Constant Bit Rate (CBR) Traffic

CBR services generate traffic at a constant rate that can be described simply by their constant bit rate. The CBR source is not bursty, and the source is active during the connection (no silent periods). Typical examples of CBR sources include voice, video, and audio (which requires more bandwidth than voice).

11.2.2 Variable Bit Rate (VBR) Traffic

In this model, traffic is alternating between two states: *on* and *off*. Traffic is generated during the *on* period at a constant rate, whereas no traffic is generated during the *off* period. The length of both periods is exponentially distributed and independent. By knowing the mean of both *on* and *off* periods, different calculations can be obtained which are useful in traffic modeling. Typical VBR sources include video-conferencing and computer network applications. The *burstiness* of a VBR traffic source is a measure of how infrequently a source sends traffic. A source that infrequently sends traffic is said to be bursty, while a source that always sends traffic at the same rate is said to be sending at a constant bit rate. The burstiness is defined in terms of the peak rate and the average rate as:

$$\text{Burstiness} = \text{peak rate}/\text{average rate}.$$

11.2.3 Pareto Traffic (Self-similar)

It was demonstrated that Ethernet and World Wide Web (WWW) traffic patterns are statistically self-similar. An on/off traffic source (VBR traffic source) with a heavy-tailed distribution, such as the Pareto distribution (discussed in Chapter 7), is self-similar.

11.3 Traffic Models for Mobile Networks

Evaluating the performance of a mobile ad hoc network depends greatly on the mobility model used. The mobility model should dictate the movement of the mobile nodes in a realistic way. Two of the mobility models mostly used by researchers are the random walk mobility model and the random waypoint mobility model. Each of these two models generates unrealistic scenarios that make them inappropriate for mobile ad hoc network simulation. An alternative is to use the Gauss–Markov mobility model to resolve the problem encountered by the previous two models.

The random walk mobility model was developed to mimic the erratic movement of entities in nature that move in unpredictable ways. A mobile node moves from one location to another by choosing two random values corresponding to speed and direction. The speed and direction are chosen to be within predefined ranges, [*speedmin*, *speedmax*] and [0, 2π], respectively. After a certain time period *t*, or a distance *d*, new values for the speed and direction are generated. No relation exists between the current and the past movements of the node. This might lead to unrealistic scenarios, where a node stops suddenly or makes sharp turns. Also, if the time period *t* or the distance *d* are small values, the node will be moving abruptly in a small region.

The random waypoint mobility model uses a pause between changes in speed and/or direction. A mobile node starts by pausing for a certain time period. Then, it moves from one location to another by choosing two random values corresponding to speed and destination. The speed is chosen to be uniformly distributed within [*minspeed*, *maxspeed*]. The mobile node travels toward the new destination at the selected speed.

On arrival, it pauses for a certain period of time and then starts the process again. The random waypoint movement is similar to the random walk movement when the pause time is zero. Hence, the random waypoint mobility model suffers from the same problems as the random walk mobility model.

To eliminate the problems encountered by these two mobility models (sudden stops and sharp turns), the Gauss–Markov mobility model is used. This model was originally proposed as a simulation of a personal communication service (PCS). However, the model has been used also for the simulation of an ad hoc network protocol. The main advantage of this model is in allowing past velocities (and directions) to influence future velocities (and directions). A mobile node starts moving using a current speed and direction. At fixed intervals of times, n, new speed and direction values are assigned to the mobile node. These values are calculated based on the values used in the previous time interval and a random variable. The speed and direction at the nth instance are given by the following equations:

$$s_n = \alpha s_{n-1} + (1-\alpha)\bar{s} + \sqrt{(1-\alpha^2)}s_{x_{n-1}}$$

and:

$$d_n = \alpha d_{n-1} + (1-\alpha)\bar{d} + \sqrt{(1-\alpha^2)}d_{x_{n-1}}$$

where s_n and d_n are the new speed and direction of the mobile node in time interval n; α, $0 \leq \alpha \leq 1$, is the tuning parameter used to vary the randomness; $\bar{s}$ and $\bar{d}$ are constants representing the mean value of speed and direction as $n \to \infty$; $s_{x_{n-1}}$ and $d_{x_{n-1}}$ are random variables from a Gaussian distribution. Totally random values (or Brownian motion) are obtained by setting $\alpha = 0$ and linear motions are obtained by setting $\alpha = 1$. Intermediate levels of randomness are obtained by varying the value of α between 0 and 1.

At each time interval, the next location is calculated based on the current location, speed, and direction of movement. Specifically, at time interval n, the mobile node's position is given by:

$$x_n = x_{n-1} + s_{n-1}\cos d_{n-1}$$

and:

$$y_n = y_{n-1} + s_{n-1}\sin d_{n-1}$$

where (x_n, y_n) and (x_{n-1}, y_{n-1}) are the x and y coordinates of the mobile node's position at the nth and $(n-1)$th time intervals, respectively, and s_{n-1} and d_{n-1} are the speed and direction of the mobile node, respectively, at the $(n-1)$th time interval.

The Gauss–Markov mobility model can eliminate the sudden stops and sharp turns encountered by the models discussed above by allowing past velocities (and directions) to influence future velocities (and directions).

Next, we discuss the most commonly used global optimization techniques to solve optimization problems.

11.4 Global Optimization Techniques

Optimization is used by operations researchers and engineers to efficiently design and evaluate the performance of an overall system. There are well-known algorithms and tools that have been developed to solve specific classes of optimization problems (e.g., linear programming, integer linear programming (ILP), and convex programming). Constrained and unconstrained optimization problems appear frequently in many engineering applications. Here, we provide an overview of some of the most commonly used global optimization techniques for solving constrained and unconstrained optimization problems and then describe the overall optimization in mathematics.

11.4.1 Genetic Algorithm

A *genetic algorithm* is a search algorithm based on the mechanics of natural selection and natural genetics using the Darwinian principle of reproduction and survival of the fittest with genetic operations. It was pioneered by John Holland as a modification of evolutionary programming in the 1960s. The idea is to construct a search algorithm modeled on the concepts of natural selection in the biological sciences. The process begins by constructing a random population of possible solutions. This population is used to create a new generation of possible solutions which is then used to create another generation of solutions, and so on. The best elements of the current generation are used to create the next generation. It is hoped that the new generation will contain "better" solutions than the previous generation and therefore give a better solution when applied to a practical problem.

11.4.2 Tabu Search

Tabu search is a heuristic procedure for solving optimization problems. It enhances the search method for finding the best possible solution of a problem. It uses a local search to repetitively move from a solution x to a solution x' in the neighborhood of x, until some conditions are reached, as shown in Figure 11.1.

The unique contribution of Tabu search is its strategy for escaping local optima in its search for better solutions. Tabu search is not best described as an algorithm, but as an

```
procedure tabusearch
begin
  select a current point, currentNode, at random
  bestNode <- currentNode
  repeat
      select a new node, newNode, that has the lowest distance in the
          neighborhood of currentNode that is not on the tabuList
      currentNode <- newNode
      if evaluation(currentNode) < evaluation(bestNode)
          bestNode <- currentNode
  until some counter reaches limit
end
```

Figure 11.1 Tabu search pseudo code (M. Guizani, A. Rayes, A. Al-Fuqaha, and G. Chaudhry, Modeling of Multimedia Traffic in High-Speed Networks, International Journal of Parallel and Distributed Systems and Networks, Vol. 3, No. 2, 2000.)

approach to solving complex problems that intelligently employs other methods to provide good solutions to the problems. Many of these were previously thought to be beyond the range of being solved.

Tabu search changes the neighborhood structure of each solution by using Tabu list. This list contains the most recent solution that has been visited through Tabu search. The attributes of solutions recently visited are labeled "Tabu-active." The process may become trapped in a local optimum space. To allow the process to search other parts of the solution space, it is required to diversify the search process, driving into new regions. This is implemented by using "frequency-based memory." The frequency information is used to penalize non-improving moves by assigning a larger penalty (frequency count adjusted by a suitable factor) to swap with greater frequency counts. This diversifying influence is allowed to operate only on those occasions when no improving moves exist. Additionally, if there is no improvement in the solution for a predefined number of iterations, the frequency information can be used for a pairwise exchange of nodes that have been explored for the least number of times in the search space, thus driving the search process into areas that have largely been unexplored so far.

The algorithm terminates if a prespecified number of iterations are reached, the objective reaches a desired value, or no improvement is achieved after a number of iterations.

11.4.3 Simulated Annealing

Simulated annealing is a generic probabilistic meta-algorithm for the global optimization problem to find a good approximation to the global optimum of a given function in a large search space. The term *simulated annealing* derives from the roughly analogous physical process of heating and then slowly cooling a substance to obtain a strong crystalline structure. The main advantages over other local search methods are that it has more flexibility and the ability to pursue global optimality. It also deals with highly nonlinear problems.

At each step, the simulated annealing algorithm considers some neighbors of the current state s in the search space that is analogous to the state of some physical system and probabilistically decides between moving the system to state s' at a lower energy or staying in the same state. The goal of simulated annealing is to bring the system from an initial state to a state with the minimum possible energy. Therefore, the simulated annealing algorithm replaces the current solution by a random solution chosen with a probability that depends on the difference between the corresponding function values and on a global parameter temperature that is gradually decreased during the process. The pseudo code for such a process is shown in Figure 11.2.

The most significant element of simulated annealing is the random number generator. It is used to generate random changes in the control variables and for the increased acceptance test while temperature decreases gradually. A good random number generator is required to tackle large-scale problems which need thousands of iterations.

```
procedure simulated annealing
begin
   initialize temperature
   select a current point, currentNode, at random
   bestNode<-currentNode
   repeat
        select a new point, newNode, randomly in the neighborhood of
currentNode
        if evaluation(newNode) < evaluation(currentNode)
            currentNode <- newNode
        else if random[0,1) < e^((evaluation(currentNode)-
evaluation(newNode))/T)
            currentNode <- newNode
        if evaluation(currentNode) < evaluation(bestNode)
            bestNode <- currentNode
        temperature <- schedule(temperature)
   until temperature < halting-criteria
end
```

Figure 11.2 Simulated annealing pseudo code (M. Guizani, A. Rayes, A. Al-Fuqaha, and G. Chaudhry, Modeling of Multimedia Traffic in High-Speed Networks, International Journal of Parallel and Distributed Systems and Networks, Vol. 3, No. 2, 2000.)

The annealing schedule determines the movements that are permitted during the search. These movements are critical to the algorithm's performance. The definition of annealing schedule is easily stated: the initial temperature should be high enough to "melt" the system completely and should be reduced toward its "freezing point" as the search progresses.

The following steps should be specified:

1. An initial temperature T_0.
2. A final temperature T_f or a stopping criterion.
3. A rule for decrementing the temperature.

During the search for the best solution, if the solution is improved, the randomly generated neighboring solution is selected. Otherwise, it is selected with a probability that depends on the extent to which it decays from the current solution.

The algorithm terminates if it reaches a specified number of iterations, or it reaches a prespecified minimum at any temperature, or there is no improvement in the solution. The maximum number of iterations is kept large.

11.5 Particle Swarm Optimization

Particle swarm optimization is a stochastic optimization technique. It bears some resemblance to evolutionary computation techniques such as genetic algorithms. The system starts with an initialized population of random solutions for optima by updating generations. However, unlike genetic algorithms, particle swarm optimization has no evolutionary operators such as crossover and mutation.

11.5.1 Solving Constrained Optimization Problems
Using Particle Swarm Optimization

Swarm intelligence is a kind of artificial intelligence based on the collective behavior of decentralized, self-organized systems. These systems are modeled by a population of agents that share information with each other and interact with their environment. Although there is no centralized control that governs how these agents will interact with each other, the local, and somehow random, interaction between these agents leads to a global system intelligence. Examples of these systems are ant colonies, flocks of birds, and schools of fish.

A number of algorithms were developed and adopted this concept in many applications. The control of unmanned vehicles in the US Army, planetary mapping at NASA, crowd simulation in movies or games, and ant-based routing are examples of these applications.

11.6 Optimization in Mathematics

Optimization refers to the study of minimizing or maximizing an objective function by finding the proper values for its variables from within the allowed set of all variables. Constrained optimization is the minimization of an objective function subject to the constraints on the values that its variables could have. In general, an optimization problem can be represented as:

$$\text{Minimize} f(x), \quad x \in S \subset R^n \tag{11.1}$$

according to the linear or nonlinear constraints:

$$g_i(x) \leq 0, \quad i = 1, \ldots, m. \tag{11.2}$$

The formulation in Equation (11.2) is not restrictive since the inequality can be represented as $-g_i(x) \geq 0$, and the constraint $g_i(x) = 0$ can be decomposed into two separate constraints, $g_i(x) \leq 0$ and $-g_i(x) \geq 0$.

Constrained optimization problems can be solved using two approaches. One is the deterministic approach, such as generalized gradient decent and feasible direction, but the drawbacks of using such methods is that they require the objective function to be continuous and differentiable, hence the ongoing research focuses more on using the other approach, such as stochastic methods like genetic algorithms, evolutionary programming, and evolutionary strategies.

11.6.1 The Penalty Approach

The most popular way to solve a constrained optimization problem is to use a *penalty function*. The search space of the problem contains two types of points, namely feasible and infeasible points. Feasible points are those which satisfy all the constraints of the problem and infeasible ones are those which violate at least one of the problem constraints.

The penalty approach addresses the optimization problem by transforming it into a sequence of unconstrained problems by adding a penalty function to the constraints on the objective function to penalize those points that violate the problem constraints. If the penalty value is high, the optimization algorithm will settle for finding a local minimum instead of finding the global minimum; and if the penalty value is small, then the algorithm will hardly discover the feasible optimal solution.

The penalty functions can be divided into two categories: stationary and non-stationary functions. In the former, a fixed penalty is added to the value of the objective function when a constraint is violated; and in the latter, a dynamically changed value for the penalty is added depending on how far the infeasible point is from the constraint. The literature shows that the results obtained are more accurate than those obtained using stationary approaches.

The penalty function is:

$$F(x) = f(x) + h(k)H(x), \quad x \in S \subset \pmb{R}^n \tag{11.3}$$

where $f(x)$ is the objective function in Equation (11.1), $h(k)$ is a dynamically changed penalty function, and k is the current iteration of the algorithm. $H(x)$ is a penalty factor defined as:

$$H(x) = \sum_{i=1}^{m} \Theta(q_i(x)) q_i(x) \hat{\gamma}(q_i(x)) \tag{11.4}$$

where $q_i(x) = \max\{0, g_i(x)\}$, $\gamma(q_i(x))$ is the power of the penalty function, $\Theta(q_i(x))$ is a multi-stage assignment function, and $g_i(x)$ are the constraints described in Equation (11.2).

11.6.2 Particle Swarm Optimization (PSO)

Particle swarm optimization (PSO) was developed as a method for optimizing continuous nonlinear mathematical functions. The ideas for this algorithm were taken from artificial intelligence, social psychology, and swarming theory. It simulates swarms of animals searching for food, like schools of fish and flocks of birds. The algorithm represents the problem by randomly creating particles which move in the solution space looking for the particle with the best solution. The algorithm relies on the concept of information sharing between particles in each program iteration in the search for the problem optima.

The ith particle of the swarm is represented by the D-dimensional vector $X_i = (x_{i1}, x_{i2}, \ldots, x_{iD})$, the particle's best position (the position at which the particle had a minimum value for the objective function) is denoted by $\pmb{P}_i = (p_{i1}, p_{i2}, \ldots, p_{iD})$, and the velocity of each particle is $\pmb{V}_i = (v_{i1}, v_{i2}, \ldots, v_{iD})$ in a D-dimensional search space.

The movement of a particle in the search space is controlled by the following movement equations:

$$V_i^{k+1} = x(wV_i^k + c_1 r_{i1}^k (P_i^k - X_i^k) + c_2 r_{i2}^k (P_g^k - X_i^k)) \tag{11.5}$$

$$X_i^{k+1} = X_i^k + V_i^{k+1} \tag{11.6}$$

where $i = 1, \ldots, N$ and N is the size of the swarm, and where x is a constriction factor which is used to control and constrict velocities, w is the inertia weight, c_1 and c_2 are two positive constants, called the cognitive and social parameter respectively, and r_{i1} and r_{i2} are random numbers uniformly distributed within the range of [0, 1].

The inertia weight w controls the effect of previous velocities of the particle on its current velocity. Large values of w facilitate the exploratory abilities of the particle while small values allow for exploitation of the local search area. Experimental results show that it is preferable to set w to a large value at the beginning of the search and then gradually decrease it when the particle approaches the best fitness value.

The relative magnitudes between $r_1 \times c_1$ and $r_2 \times c_2$ determine whether the particle moves toward *pBest* or *gBest*. If the upper bound of $r_1 \times c_1$ is greater than the upper bound of $r_2 \times c_2$, then the particle tends to utilize the neighborhood experience more than its own experience in finding a better fitness value. The values c_1 and c_2 are generated randomly for each particle at each iteration so that the particles may vary their influence among the different sources of information.

11.6.3 The Algorithm

The algorithm for PSO can be written as follows:

```
For each particle
{
  Do
  {
    Initialize particle
  }
}
  Do
  {
    For each particle
    {
      Calculate the corresponding fitness value
      If the fitness value is better than the particle's best
        fitness value
        Set the current P vector to the particle's current X
          vector
    }
  Choose the particle with the lowest fitness value and make it
    the global best position
    For each particle
    {
      Calculate the particle's velocity according to Equation
        (11.5)
      Update the particle's current position vector X
    }
  } while maximum iteration or minimum error criteria is not
    attained
```

The algorithm starts by creating the swarm of particles and assigning to each particle its parameters, e.g., its initial position. Then it updates the position of each particle according to Equations (11.5) and (11.6). At each iteration, the particle compares its current position to its best ever position. If the current position is better than that position, the current position becomes the particle's position. The particle with the best value for the fitness function is chosen to be the swarm's best particle and the particles in the swarm tend to fly toward this particle.

The main advantages of the PSO are:

1. It does not have many parameters to tune to in order to get an acceptable performance.
2. It is applicable for both constrained and unconstrained problems.
3. It is easy to implement.

On the other hand, the PSO can converge prematurely if the penalty values are high and will be trapped in local optima. Also, the parameters are problem dependent and finding the best values for those parameters is not trivial.

11.7 Summary

In this chapter we studied different traffic models used to simulate network traffic loads. The models were divided into two main categories: models that exhibit long-range dependencies or self-similarities; and Markov models that exhibit only short-range dependence. Self-similar models are typically used to model non-homogeneous traffic loads, whereas Markov models are used to model traditional networks such as PSTN.

In the chapter we also introduced the most common optimization algorithms and tools that have been developed to solve specific classes of optimization problems. We provided an overview of a few global optimization techniques for solving constrained and unconstrained optimization problems and then described the overall optimization in mathematics.

Recommended Reading

[1] J. Banks, *Discrete-Event System Simulation*, 3rd edition, Prentice Hall, 2000.
[2] J. Broch, D. A. Maltz, D. B. Johnson, Y.-C. Hu, and J. Jetcheva, "A performance comparison of multi-hop wireless ad hoc network routing protocols," *Proceedings of MobiCom, the ACM International Conference on Mobile Computing and Networking, Dallas, USA*, October 1998.
[3] T. Camp, J. Boleng, and V. Davies, "A survey of mobility models for ad hoc network research," *Journal of Communication & Mobile Computing (WCMC): Special issue on Mobile Ad Hoc Networking: Research, Trends and Applications*, vol. 2, no. 5, pp. 483–502, 2002.

[4] D. Johnson and D. Maltz, "Dynamic source routing in ad hoc wireless networks," in *Mobile Computing* (T. Imelinsky and H. Korth, eds.), pp. 153–181, Kluwer Academic, 1996.

[5] A. M. Law and W. D. Kelton, *Simulation Modeling and Analysis*, 3rd Edition, McGraw-Hill, 2000.

[6] B. Liang and Z. Haas, "Predictive distance-based mobility management for PCS networks," *Proceedings of the Joint Conference of the IEEE Computer and Communications Societies (INFOCOM)*, March 1999.

[7] M. Zukerman, T. D. Neame, and R.G. Addie, "Internet traffic modeling and future technology implications," *Proceedings of INFOCOM*, 2003.

[8] D. E. McDysan and D. L. Sophn, *ATM: Theory and Application*, McGraw-Hill, 1994.

[9] R. O. Onvural, *Asynchronous Transfer Mode Networks: Performance Issues*, 2nd Edition, Artech House, 1995.

[10] R. Guérin, H. Ahmadi, and M. Naghshineh, "Equivalent capacity and its application to bandwidth allocation in high-speed networks," *IEEE Journal on Selected Areas in Communications*, vol. 9, no. 7, pp. 968–981, 1991.

[11] W. Willinger, M. Taqqu, R. Sherman, and D. Wilson, "Self-similarity through high variability: statistical analysis of Ethernet LAN traffic at the source level," *IEEE/ACM Transactions on Networking*, vol. 5, no. 1, pp. 71–86, 1997.

[12] M. Guizani and A. Rayes, *Designing ATM Switching Networks*, McGraw-Hill, 1997.

[13] J. Kennedy and R. Eberhart, "Particle swarm optimization," *Proceedings of the IEEE International Conference on Neural Networks*, IEEE, 1995, vol. 4, pp. 1942–1948.

[14] K. Parsopoulos and M. Vrahatis, "Particle swarm optimization method for constrained optimization problems," *Frontiers in Artificial Intelligence and Applications*, vol. 76, pp. 215–220, 2002.

[15] X. Hu and R. C. Eberhart, "Solving constrained nonlinear optimization problems with particle swarm optimi-zation," *IEEE Conference on Cybernetics and Intelligent Systems*, pp. 1–7, June 2006.

Index

Network Modeling and Simulation M. Guizani, A. Rayes, B. Khan and A. Al-Fuqaha
© 2010 John Wiley & Sons, Ltd.